NOTHING
BUT THE
TRUTH

NOTHING
BUT THE
TRUTH

LARRY RHODES

ISBN: 978-1-959483-96-0 (sc)
ISBN: 978-1-963068-06-1 (sc)

Library of Congress Control Number: 2023923599

Contents

CHAPTER 1

Ross Daniels could be a poster boy for the Southern California lifestyle. Blond hair, blue eyes, tanned and athletic, Ross had the fortune to be born the son of a wealthy financier. He was no dumb blond, however, as evidenced by the Phi Beta Kappa key on his keychain. Having graduated with high honors from the University of Southern California, Ross had been admitted to post-graduate studies in political science. His ambitions included no less than a seat in the State House of Representatives or Senate or perhaps the U.S. Congress.

Ross had spent the first portion of the summer break relaxing in Europe, returning just in time for a big Independence Day party to be hosted by his closest friend, Bill Kennedy. Bill and Ross had become fast friends in high school and although he lacked Ross' good looks, Bill was already a legend as a ladies man and society mainstay. He was the son of an even wealthier media mogul who owned a string of newspapers, radio and TV stations. As soon as he completed a degree in business, his father had turned over the operation of sizable charitable foundation. Bill was set for life, and he hadn't even celebrated his 22nd birthday.

An early evening chill in Beverly Hills had forced Ross Daniels to roll up the windows on his Ferrari Spider convertible, but he had not put the top up – only rain would force him to do that. The Spider had been an early college graduation gift, and Ross had already put over 50,000 miles on it in less than six months. He had quickly conquered the twisting canyon roads and treacherous hairpin turns on the road between USC and his home.

Independence Day fireworks lit the sky as Adam Daniels turned into the private drive that led to Bill Kennedy's mansion. His friends were still talking about last year's 4th of July party. The invitation-only feast had been filled with "B" list movie stars and up-and-coming rock musicians. Last year's brawl and police raid had even made the late news on local TV stations and the <u>LA Times</u>. Even though Bill's father owned several TV and radio stations, he could not keep the fracas out of the media. This year's party would undoubtedly be more constrained.

Tom Perkins immediately recognized Daniels and quickly opened the guard gate. Adam slowed as he approached the gate and suddenly remembered the manila envelope on the passenger seat. He quickly stuffed it under the seat and lowered his window. "Good evening, Tom."

"Good evening, Mr. Daniels. Quite a crowd tonight."

"No complaints, I hope."

"No, sir. Not yet anyway."

Adam waved and entered the large circular drive in front of the mansion. It was relatively early, but cars were already spilling out onto the lawn. Adam spied a spot right in front and quickly maneuvered his Ferrari into the space. It was a cool evening and he started to put the top on the convertible but was distracted by a bevy of beautiful girls entering the party. He pulled the envelope from the floor of the car and quickly stuffed it into his jacket. Adam was used to having money, and being around expensive things but he didn't like carrying large amounts of cash around. Bill had asked him to take some money from a lock box at a bank they shared.

"Action News" Reporter Daniel Turner straightened his tie and waited for the signal from his cameraman.

"Good Afternoon. This is Daniel Turner reporting live from the 228nd district courtroom where a short while ago, the jury in The State of Texas vs. John Landrew murder trial sent a message to Judge Thomas, that after two days of deliberations, they were hopelessly deadlocked at six for acquittal and six guilty. The verdict produced deep emotions on both sides, as the family of the defendant erupted in cries of disbelief and anger at the prospect of starting all over again. Assistant District

Attorney Phil Conley held a short news conference after the court was adjourned. He expressed his frustration at the decision but thanked the jury for their quick work. He said he would begin preparation for a new trial immediately. The family of the defendant and the defense team had no comment on the decision and left the courtroom quickly."

A stony silence greeted Phil Conley as he entered the District Attorney's offices. Several staff members attempted to console him and wished him luck in the new trial.

Law clerk Tom Benning gave him a 'head's up'. "The big guy's waiting in your office."

"Thanks for the warning, Tom."

A burly six foot four inches and two hundred and fifty pounds, most people would assume that District Attorney Bill Watson was a professional wrestler.

June Smith, the department secretary pushed a phone message into his hand. "Phil, Allen Atkins called. He said he tried leaving messages on your phone but as a backup wanted me to give you this in case you came back to the office. He asked if you could meet him tonight around 6:00PM at 'Jake's Bar' in Clear Lake."

Phil nodded in response as he took a deep breath and opened the door to his office. Watson was holding a file folder and staring out the window. He glanced at Phil and sat in the nearest chair. "I thought you said we had this one for sure. What happened?"

Phil pulled a chair over to Watson. "It's just another one of those cases where the jury decides to ignore the evidence."

Watson threw the folder on Phil's desk. "Almost two million dollars and a re-trial will probably be more than that."

Phil shrugged. "I know Landrew's alibi is lying, but I just can't prove it."

Watson stood up. "I won't agree to a retrial until you can." He turned and stalked out of the office, slamming the door. Phil sighed and looked at the note in his hand.

Phil abhorred being late for anything, but the Gulf Freeway traffic from the DA's office in the downtown area of Houston to Clear Lake City had been a virtual parking lot. He spied Ann Stevens at the bar and waved as he wove his way through the early evening crowd of yuccies and business professionals.

He was impressed. He hadn't been to "Jake's Milky Way Bar" in a while. He realized they had spent quite a bit of money remodeling it into an art deco and outer space theme. The NASA space center was only a few miles down the road and most of the early evening crowd worked there or worked for contracting companies that worked almost exclusively for NASA.

He approached Ann slowly and cautiously held out his hand. She hugged him and kissed him on the lips before he could duck out of the way. "How's the family?" he asked her. He hadn't seen very much of Ann in the last six months, as he had often worked 12-to-14-hour days on the murder case.

"Dan's away as usual. The kids are getting bigger and more demanding than ever." She returned to her bar stool. "I just heard about the verdict. It's a real bummer. I know your team worked like mad on it."

"Thanks. We thought we had a good case. The last thing we expected was the jury to deadlock on it." He ordered a drink and looked around. "Where's Allen? I thought I was late because of the traffic."

"He called me a minute ago. He's running late too but said he would be here in a few minutes."

Phil settled onto a stool, leaving an empty stool for Allen, but Ann moved over next to him, bumping into him on purpose. Phil started to move to the next stool, but Ann put her hand on his. "I won't bite, Phil." She then muttered under her breath, "Not in public, anyway."

Phil smiled nervously. "So, what's up with him? What's so urgent?" The bartender delivered his drink and he began to sip on it slowly, as he hadn't eaten in quite some time.

"I don't know, he said he wants to show us something, but didn't say what."

Always the business professional, Ann was wearing a dark, expensive business suit and her usual "Modern Princess Diana" bob haircut. Her blue eyes contrasted sharply with her almost black hair. He knew Ann exercised routinely to maintain her figure, and if they hadn't gone to high school together, Phil might have guessed she was younger than 33 years.

She did seem a little tired though. He wondered if Ann were under some work-related stress.

"How's the medical equipment business?"

"Busy as hell! We're coming out with a new heart monitor for ICUs that we hope can predict an attack based on changes in rhythms."

"That's great! I heard you're thinking of staffing up some." Phil began nibbling on some bar munchies.

Phil kept glancing past Ann, piquing her curiosity. She turned slowly and saw two beautiful young women at the end of the bar, conversing and sharing drinks with an expensively dressed man old enough to be their father. One of them was slyly flirting with Phil.

Ann was frowning at him when he looked back at her. "Where did you hear about the staffing increase? We haven't told anyone outside the company yet!"

Phil smirked. "Through a friend of a friend. Sort of a grapevine thing."

Phil motioned to the bartender, and Ann watched as Phil leaned over and said something to him. The bartender looked at the two pretty girls and nodded. A moment later, he delivered some drinks to them, indicating they were from Phil. One woman blew a kiss to Phil and the other stuffed a napkin with her phone number into an empty glass and slid it to Phil. Ann shook her head in disbelief as she watched Phil enter the phone number in his phone's contact list.

"They're too young for you, Phil. Oh... here's Allen."

Allan Atkins entered the bar carrying a small briefcase. At six feet even, Allen was a few inches shorter than Phil, but he maintained a daily exercise regimen that kept him trim. Phil's lack of exercise forced him to maintain a strict diet to avoid the curse of the middle age spread. Allen's dark hair and eyes also contrasted sharply with Phil's blond hair and hazel eyes. As he drew near, Ann and Phil snickered as they noticed his NASA cap, NASA shorts, and NASA T-shirt.

"So... Allen. Do you work for NASA?"

Allen ignored Phil's remark and set his briefcase on the bar. "Department picnic. How are you guys?" He shook Phil's hand and Ann hugged him warmly.

"Okay! How's work?"

"Busy, Ann, but I love it that way." He settled onto an empty stool next to Phil. "I heard about the verdict. I'm sure you'll get him the next time around."

Phil nodded. "He's guilty of at least second-degree murder. I'm going to do everything I can to make sure he'll be in prison for a long time." Phil finished his drink. "But I don't think you asked us here to talk about that. Have you completed the property settlement with Mary or something?"

They could sense this was still a sore point with Allen. He gritted his teeth and struggled to answer in a normal tone. "Not yet."

"Sorry to hear that." Phil put his hand on Allen's shoulder. "How are you getting along?"

"Okay, I guess. I have some good days and some bad days. I seem to be spending even more time in my lab than I used to. I guess it helps me forget I don't have a wife anymore."

The three long time friends sat for a moment quietly sipping on their drinks. Allen finally broke the silence. "Phil, do you remember how drunk we got the night my divorce became final?"

Phil laughed. "Not really, but I do remember how you whined about the way people can lie and how often they get away with it."

"Of course you do. I haven't stopped complaining about it. If I could have proved she was lying, she wouldn't have won.... Well, she might have won anyway, but she wouldn't have gotten away with accusing me of spousal abuse on a public record, which was a pack of lies."

While Ann was sympathetic to Allen's problem she also wondered where this conversation was leading. She glanced at her watch and realized she needed to leave soon for a business dinner meeting. "What's going on Allen?"

"Well, the reason I wanted to talk to you is... I think I've come up with a nearly fool proof lie detector."

Ann and Phil looked at each other and laughed.

Phil signaled the bartender for another drink. "Come on, there is no such thing"

"Just a second." Allen opened the briefcase, took out a small black box, sat it on the bar near Phil, and put the briefcase on the floor.

Phil picked it up and studied it. "What's this?"

"That's a Voice Stress Analyzer or VSA for short. This one is probably the best commercial model made." He took a quick sip of his drink. "I've made some software modifications, of course."

Phil handed the VSA to Ann and frowned at Allen. "Voice Stress? Doesn't that have something to do with changes in the vocal cords when someone is lying?"

"Exactly right. When someone lies, their muscles tighten up some, including the muscles in their voice box." He pointed to a small silver disk on the top of the VSA. "This microphone picks up something called 'micro tremors' that happen a few thousands-of-a-second after someone tells a lie."

The VSA in Ann's hand was only a little larger than a pack of cigarettes. "I thought these things weren't very reliable?"

Phil watched Ann hand the VSA back to Allen. "I know they are not reliable enough to be used in a court of law. Federal laws even prohibit most uses of lie detectors in the workplace. You can't even make a prospective employee submit to one, for example, so this can't be your foolproof lie detector."

"No. I just brought it to demonstrate a principle. My method combines this technique and several others and then uses an Artificial Intelligence algorithm to produce a probability of truthfulness."

Phil wasn't buying it. "Are you bullshitting us?" He looked at Ann. "Is he kidding?"

Ann shrugged. "Who knows?"

"Wait! Let me demonstrate a principle." Allen was sitting next to an attractive, conservatively dressed woman that he guessed was probably thirty or so. She appeared to be a natural redhead with shoulder length curly hair. A yuccie barely old enough to be in the bar was trying to get her to go to dinner with him and had almost succeeded when he excused himself to go the bathroom. She glanced at Allen with the clearest, prettiest blue eyes he had ever seen. She wasn't wearing a wedding band but, from his experience, that didn't necessarily mean anything. He would have to be clever, yet convincing.

"Hi. My name is Allen." He smiled at her and to his surprise, she answered him.

"Hi. I'm Joanne." She glanced past him at Phil and Ann.

Allen, gestured at his friends. "This is Phil and Ann."

She smiled at them. "Hi."

"Joanne, would you like to participate in an experiment?"

She took a sip on her drink and glanced at the NASA symbols on Allen's clothes. "Experiment? You mean like a survey?"

"Not exactly." Allen put the small VSA on the bar in front of her and pressed a button on top. It beeped several times, and a red light flashed a few times and went out. "This is an experimental lie detector. When your friend returns, why don't you see if he's for real?"

A shocked expression ran over her face and then disappeared. "What?"

"You could ask him some pointed questions. If he's lying, the little red light on top will flash."

"I can't do that! It's probably illegal or something." She glanced furtively around the bar as if afraid someone had heard them.

Phil reassured her. "As a lawyer, I can assure you it's not."

Allen egged her on. "Why don't you give it a try? Just put your purse in front of it so he won't see it."

"You better decide soon," Ann urged. "He's coming back."

Joanne looked at the small black box and then at Allen. Just before the yuccie returned, she put her purse on the bar so he couldn't see it.

"So, Joanne, where do you want to go to dinner?"

She looked at him for a moment. What should she do? She swallowed hard.

"Oh... Mike. I'm not sure I want to... now."

Mike was surprised but not deterred. He pushed his stool closer to hers and sat down. "Are you sure? Why did you change your mind?"

"I guess I got to thinking... why would a good looking, really young guy like you want to have dinner with an older woman?"

Phil and Ann were pretending to watch a ball game on a TV behind the bar. Allen had to take a deep drink and Phil coughed to keep from laughing. Ann pretended to sneeze.

"What? Why wouldn't I? You are an extremely attractive woman." Mike was upset now, and it showed.

The little red light on the top of the VSA flashed briefly. Joanne noticed as did Allen, Phil and Ann. She looked at Allen's reflection in the mirror behind the bar as he smiled at her.

The lie touched a sore spot in Joanne and a deep-seated anger popped to the surface.

"Maybe you should just go home to your wife!" She snapped as she tried hard to calm down.

"Wife? I'm not married!" He croaked. The expression on his face convinced Joanne she had caught him in yet another lie. She also noticed the red light blinking on the VSA. She turned to him angrily.

"I'd rather eat alone than with...." Her voice trailed off as she bit her lip.

"Than with what? What's going on here?" Mike looked past her at Allen, Ann and Phil but they were looking off elsewhere, at their drinks, at the TV, at the pretty girls at the other end of the bar.

"Just go home!"

"What are you talking about, Joanne? I told you I'm not married!"

The red light blinked again. That was enough for Joanne. She turned to him and yelled. "Try that line on some other fool!"

Mike saw the other patrons looking at them as Joanne took a drink. She was fuming as she glared at the bartender, who grinned. Mike huffed and puffed, finished a drink on the bar, left some money and started to walk off. He stopped to glance back and then left with a storm cloud over his head.

"I'm sorry. Maybe that wasn't such a good idea." Allen looked embarrassed as he spoke softly to her. Maybe interfering in other people's lives was not the best way to demonstrate an idea, even a good one.

"No, it was a great idea! He was really convincing. I'm sure he had only one thing on his mind." She picked up the VSA and started examining it.

"Could I buy one of these?"

Ann, Allen, Phil, and the bartender laughed loudly.

Allen held his hand out to her. "My name is Allen Atkins."

She shook his hand. "Joanne Davidson. It's nice to meet you." She held the VSA out toward Allen. "Are you married, Allen?"

They all laughed again.

"Recently divorced."

She pointed the VSA at Phil. "How about you, Phil?"

Phil held his hands up as if she were pointing a gun at him. "Never married."

Joanne pointed it at the bartender who had been listening intently to the conversation. He smiled and leaned forward as if speaking into a microphone. "I'm available, honey."

The red light didn't blink, but Joanne observed that he was wearing an earring, had a beer belly and a large tattoo on one arm, was slightly balding and had quite a bit of gray in a scraggly beard.

Not my type! She gave him a dirty look and started to point it at Ann but changed her mind.

Ann chuckled softly. "Separated."

Allen and Phil both looked at Ann with surprised expressions.

Phil put his hand on her shoulder. "I thought you and Dan had gotten back together."

Ann shook her head. "Looks like were heading for the big D."

"I'm sorry, Ann. I didn't know," Allen said sympathetically.

The light on the VSA had not blinked. Allen remembered Joanne. "How about you, Joanne?"

"Not married."

"Boy friend? Fiancé?" he asked, half seriously.

"Not at the present time," she chuckled. "How about you? Girl friend? Fiancée?"

Allen laughed. "No, to both."

The light didn't blink, and Joanne studied the VSA carefully.

"My girlfriends and I could really use one of these. Where could we get one?"

"That's a pretty expensive toy. I've made some improvements, but it's not foolproof."

"Maybe not, but if it could eliminate lies there would be a lot less tears... and heartaches."

As Joanne continued to examine the VSA, Allen looked at Phil. "Actually, I have a better system."

"When could we see the prototype?"

Allen finished his drink and waved off the bartender who motioned him for another. "How about tomorrow morning? Say, around 9:00AM or so?"

"That would be perfect. I'll be there for sure." Ann was already entering the time in her phone's calendar.

"Count me in, too!" Phil watched Joanne examine the VSA.

"Great, I'll leave word with the receptionist."

Ann looked at her watch. "I have a business dinner to go to. I'll see you guys tomorrow. Nice to meet you, Joanne." Ann shook hands with Joanne and patted Phil and Allen on the shoulder as she passed them.

Allen looked at his watch. "Oh... I have to go too. My brother and his family are coming tonight to spend a few days. Nice to meet you, Joanne." He shook her hand and picked the VSA up, placed it gently in the foam-lined briefcase and walked quickly out.

Joanne and Phil regarded each other for a moment. When he looked away for the bartender, she muttered under her breath. "I wish I still had that monitor." Phil didn't hear her as he was indicating that he wanted the check.

"So, Phil, do you have a girl friend or fiancée?"

Phil smiled slowly. "No." He decided to change the subject. "Do you live in Clear Lake?"

"No, I'm attending a medical convention here. I work in the Medical Center... as an ER nurse." Phil had been standing back from the bar and when he sat down in a more lighted area, Joanne frowned. *I've seen him somewhere!* She began studying his face.

Phil finished his drink. "ER nursing is pretty stressful, isn't it?"

"You get used to it. I'll bet trials can be stressful too."

Phil shrugged his shoulders. "Only while you're waiting on the verdict."

The murder trial! "You were the prosecutor on that murder trial with the hung jury!"

Phil nodded unhappily. *Why do people only remember the ones you lose?*

Joanne finished her drink. "Allen obviously works for NASA. How did you meet him?"

"Ann, Allen and I went to high school together. Ann got a degree in business; Allen went to MIT to study physics and I... went to law school...."

Joanne was gazing into his eyes and he lost his train of thought. *What pretty eyes!*

"I sure wish I had his lie detector, Phil. You know, life is strange sometimes."

"What do you mean?"

"Well, if it wasn't for a lie, I wouldn't have met you."

Phil thought about that and smiled. He put his hand on hers and she put her other hand on his.

"Would you like to have dinner with me?"

Joanne's eyes twinkled. "I would love to Phil."

"I promise I don't have an ulterior motive."

"Oh... darn."

They laughed. Phil paid the tab as Joanne collected her purse.

CHAPTER 2

Janet Turner was pretty and petite. She had previously worked for a NASA subcontractor and had caught the attention of several scientists and engineers who had recommended her for a permanent position. She had held several positions in the computer science department in three years and currently was a research analyst.

Allen was shy by nature and his intense attraction to her had almost rendered him mute when she was around. He had planned to ask her out numerous times since his divorce, only to chicken out at the last moment. Through his department secretary, he knew Janet had been married before and divorced, but it didn't appear that Janet currently had a boyfriend. Allen had been too shy to ask his source. If he only KNEW she wasn't involved with someone, maybe... When Allen finished the lie detector prototype, he tested it on the night janitor who thought it was a big joke. Allen suddenly realized he had an opportunity to determine Janet's availability. He was surprised when she accepted his invitation to see his lab.

Janet Turner guessed that Allen liked her. He tried not to show it, but she could tell. She sort of liked him, but he almost never spoke when he was around her, so she couldn't be sure. When he asked her if she would like to see some new equipment he was working on, she quickly accepted. She knew he was working on some long-term projects and she was curious.

Allen's lab was very clean and orderly. She recognized some equipment from several laboratory classes she had taken in her upperclassman days in college. Allen walked right past it and opened a door to a small lab in the back of the main lab. There was so much equipment in the room she wondered how it all fit in there. Most of the equipment was mounted on racks around the walls, but there was a table in the middle with a computer, a video camera and several other items she didn't recognize.

Two ghostly images appeared in a window on the computer monitor and a "status" window opened up. A message appeared on the monitor that was logged to a "lab activity" file.

08:05 AM LAB ACTIVITY FILE OPENED, TWO PERSONS IN RANGE.

Janet stared in amazement at the equipment crammed into the small lab. "So, this is where you spend most of your time. It's kind of cramped in here. Why don't you set all this stuff up in the main lab? There's plenty of room."

"I don't need a lot of room, and nobody bothers me in here. Have a seat, please." As she settled into a chair, he remembered his manners. "Would you like a cup of coffee?"

Another message flashed on the computer screen.

08:06 AM ALLEN ATKINS IDENTIFIED BY VOICEPRINT ANALYSIS. OTHER PERSON UNKNOWN.

"Yes, please, just cream."

"I'll be right back."

When Allen walked out, Janet stood up and walked around the room examining the equipment. Allen returned with coffee for both of them, set hers carefully on the table and casually sat down in front of the computer.

"Have a seat, Janet."

"What do you do here, Allen?" She sat down precisely where Allen hoped she would. He also hoped she didn't notice one of the small boxes on the table rotating on a motorized mount.

"We are developing some detectors that will monitor the breathing of astronauts during their sleep on long-duration missions."

"Really! That's pretty far into the future, isn't it?"

"Probably, but the same methods could work on monkeys, and we may be sending them on long missions in a few years. I am also working on several other projects in the early planning phase." He pointed to one of the boxes on the table. "This box is a highly miniaturized spectrophotometer. It measures very low levels of carbon dioxide and water vapor. We could use a laser to excite the molecules in an astronaut's breath and measure the concentration of several other gases. This could give an indication of his metabolic state." She was smiling at him, and Allen realized he was talking excitedly and might be losing her. "Oh... sorry. I get carried away sometimes." Another window opened on the screen, and he glanced at its message.

08:08 AM INFRARED ANALYSIS COMMENCING ON UNKNOWN PERSON.

08:08 AM SPECTROSCOPIC ANALYSIS COMMENCING ON UNKNOWN PERSON.

"It's okay, Allen. It's nice sometimes to talk to people who are excited about their work. I don't think that's ever happened to me. I like what I do, but it's not a passion for me."

08:09 AM DATA ANALYSIS FUNCTION ACTIVATED.

08:09 AM PLEASE ENTER SUBJECT'S NAME....

As Janet leaned over to look at the box Allen had pointed to, she didn't notice him typing her name on the computer's keyboard. Allen watched several more windows open on the computer screen as Janet looked at some of the other equipment in the room.

08:10 AM VOICE STRESS ANALYSIS ACTIVE.

08:10AM AI ALGORITHM ACTIVE.

08:11AM ENTER DATA ANALYSIS TYPE....

Allen moved a mouse pointer to the "data analysis type" window and clicked the mouse. He scrolled down a list and selected "Data Gathering" and "Interview."

"Pretty impressive, Allen."

After a moment, she sat back and folded her arms. "But why did you really ask me to come here today? You know I don't know very much about this stuff."

Allen blushed, and Janet sensed she had caught him by surprise.

"You're right. It was the only way I could get you alone to ask you if you would like to go out to dinner with me on Saturday night. Maybe even see a movie?"

Janet smiled and leaned forward. "I thought so. You didn't have to get me alone to ask. Actually though, my parents are coming this weekend. Maybe another weekend?"

Allen was excited. *Some progress!*

"How about next weekend?"

"I'm going to a high school reunion that weekend, or I would say yes."

"How about dinner one night this week?"

"I'm really busy this week. I'm studying for final exams. You know I'm working on a master's degree."

This isn't working. "It sounds like you are putting me off. Maybe there won't be another weekend available."

"Now, Allen, don't act that way. Look, I'll call you tomorrow and we'll set something up for one day next week."

"Sure." Allen tried to hide the disappointment in his voice.

"Well, I have to go. I'm almost late for a meeting."

Allen stood up as she paused at the door.

"Who knows? I may even invite you over for a home-cooked meal. I'm a pretty good cook."

He grinned. "I would really like that."

She winked at him as she opened the door. "See you...."

Allen sat down at the computer to think about their encounter. "Yes!!!" he shouted at the computer. The computer beeped back at him. He sat looking at the screen for a moment, then thought of something Joanne had said.

"No lies... no tears... no heartaches."

He scrolled through the considerable text the computer's voice-recognition system had recorded and corrected several words. He typed in a "Run Analysis" command and waited for the results. When it was finished, he turned on the computer's speakers, sat back and folded his arms. "Sherlock?"

A deep, synthesized voice answered him. "Yes, Allen?"

"So... did Janet respond truthfully to the questions I asked?"

"Yes. All responses were truthful."

"She didn't say no when I asked her out."

"I'm not sure what you mean. Please restate your comment as a question."

"She didn't say she had a boyfriend."

"I don't understand. There was no question about a boyfriend."

Allen rubbed his chin, thinking of Janet.

"What is the probability that Janet will go out with me?"

Another window popped open on the screen as Sherlock checked the meaning of boyfriend. "Based on questions one and two, the probability is sixty percent. But it is based on a very limited set of data, of course."

"Well, that's a little better than even odds," Allen mused.

"I didn't understand your last comment. Please restate it as a question."

"Never mind. Thank you, Sherlock."

"Quite all right, Allen."

Allen chuckled as he turned the computer speakers off.

Chapter 3

Allen's lab had recently been remodeled and Ann and Phil were impressed with its new, bright, and modern look. There was a lot of electronic equipment on racks and even more sitting on tables, but Phil and Ann couldn't recognize any of it. Phil saw Allen in the back of the lab wearing a white lab coat and dark sunglasses, tinkering with a large table-mounted instrument. Both Phil and Ann immediately noticed the "Danger, Laser Light" sign on the front.

"So, where is this foolproof lie detector, Allen?"

"Oh, hi Ann. Right on time. Follow me."

He took off his sunglasses and led them to the door in the back of the lab that led to the small room full of electronic gear. Their attention was immediately drawn to a table in the middle that held a computer, a video camera, and several other strange looking items.

"Have a seat, guys."

As Phil, Ann and Allen entered the lab, several motion and infrared sensors recorded their entry. A status window opened up on the computer monitor, and a memo from "Sherlock" flashed on the screen:

09:05AM LAB ACTIVITY FILE OPENED. THREE PERSONS IN RANGE.

Phil and Ann looked around the lab in amazement at a dazzling array of wall-mounted equipment stuffed into the room.

"Where is the prototype, Allen?"

"It's all around you, Phil."

"It's ALL of this!" Phil and Ann looked at each other in disbelief.

"Well, not all of it. Most of it is now in this computer and these boxes on the table."

Allen motioned them to sit down. Ann unknowingly sat in the "subject's" chair while Phil walked around the room. The status window on the computer screen flashed a new message, as if in response to an unasked question.

09:06AM ALLEN ATKINS IDENTIFIED VIA VOICEPRINT. TWO UNKNOWN PERSONS.

Allen sat down at the computer, glanced at the screen, and smiled. He pressed a few keys, folded his arms, and sat back in his chair with an amused expression as Ann and Phil continued to look at the equipment around them. Neither noticed the small metal box rotating slowly on a motorized mount. The computer was re-aiming a carbon dioxide laser to point to the space in front of Ann's mouth.

Two new windows opened on the computer screen: one labeled 'Infrared Imaging'; the other 'Spectroscopic Analysis'. A label on the bottom of each window changed from 'inactive' to 'active'. Allen watched as an image of Ann's head appeared on the screen in a myriad of colors, and her breath appeared on the screen in a kaleidoscope of colors. He glanced at the status window on the computer's screen.

09:08AM INFRARED ANALYSIS COMMENCING ON UNKNOWN PERSON.

09:09AM SPECTROSCOPIC ANALYSIS COMMENCING ON UNKNOWN PERSON.

09:09AM VOICE STRESS ANALYSIS ACTIVE.

09:09AM AI ALGORITHM ACTIVE.

Phil stared at the screen for a moment. "How does it work?"

Allen began his explanation by pointing to a sleek unidirectional microphone on the table aimed at Ann. "This microphone is super sensitive. It could pick up what you're saying in a crowd at 30 feet. The voice data is then interpreted by voice-recognition software and a voice stress analysis program."

"Is that similar to the VSA you showed us last night?"

"Yes, Ann, but this software utilizes a more sophisticated model than the hand held VSA and quite a bit of reference data is stored in its database. Generally speaking, the more data, the more accurate the analysis." Allen the pointed to one of two small metal boxes on the table.

"This is an infrared camera. It assigns different temperatures different colors. These cameras typically help engineers and scientists find hot spots or cold spots in equipment that may affect their operation or that need to be insulated."

Allen turned to a large computer monitor mounted on a rack behind him and switched it on.

After a moment, a multi-colored image of Ann's head appeared on the monitor.

"This is your head, Ann."

Most of Ann's image was dark with a few lighter colors on various spots on her face.

Phil saw a small window pop open on the computer screen with the same image as the large monitor. "What's that used for?"

"I have a theory that temperature profile patterns develop on the face and head with changes in blood flow. I'm hoping the patterns will be different when someone is lying than when they are telling the truth and, with sufficient data, the Artificial Intelligence algorithm will be able to tell the difference."

"Sounds like quite a long shot."

"I agree, Ann. By itself, it probably wouldn't be very reliable, but combined with other techniques, it may help in the overall analysis."

Phil pointed to the other metal box on the table. "What's that other thing?"

Allen patted the small box gently. "This is sort of the heart of the system. It's a miniaturized spectrophotometer and a carbon dioxide laser. When laser energy passes through a gas, it gives up some of its energy by exciting it. Certain gases relax back to their normal state by emitting the excess energy back as light in the infrared region. This is known as florescence. This spectrophotometer looks at the florescence of some gases at several wavelengths and measures their intensity. The change in the intensity of these gases from one breath to the next is the information processed by the AI algorithm along with the Voice Stress Analysis and the pattern of hot and cold areas on a subject's head."

"And the AI Algorithm puts all this together?" Skepticism was written on Phil's face.

"Yes."

"What happens then?"

"A special program correlates the data and assigns a probability value of truthfulness to a response, which ranges from zero to one hundred percent."

"How is this system different from a regular polygraph?"

"People can fool most lie detectors because they can be trained to show no emotion or take drugs in advance of a test. With this system there are no visible signs they are even being monitored."

Ann leaned across the table staring at the small spectrophotometer. "What about the laser beam? Wouldn't that give it away?"

"A carbon dioxide laser's beam is invisible to the human eye."

Phil seemed to be thinking about the possible advantages of an "invisible" lie detector. Ann fidgeted in her chair. "When can we try it?"

"Right now. But first, I'd like to tell you about another aspect of the AI algorithm software running here." Phil and Ann waited for him to begin.

"I had a roommate at MIT, Tom Kelley, who was a real computer nerd. I once caught him talking to one and thought he had finally flipped out, but he explained he was working on a voice-recognition program. A few years ago, I talked him into helping me adapt an AI algorithm to help solve crimes."

Phil frowned. "There are several commercially available crime-solving programs on the market, and the police departments of large cities sometimes use three or more to help them solve difficult cases."

"I know that, but this one is a little different. Tom's wife is a speech therapist and she helped us develop a natural language interface to the AI software system. With her help we enabled it to ask questions when there was a conflict in the data entered, or when it determined that some area needed further clarification."

"What?" exclaimed Phil. "The computer could ask questions? What happened to this software?"

Allen laughed. "We formed a software company to sell it. Tom still works on it every now and then entering data from cases that were solved to help give it a database of knowledge. He sells a few systems a year. But he has a day job to pay the rent."

"How come you never told us about this before?" asked Phil.

Allen shrugged. "I guess it just never came up."

Ann was making notes in her phone. "What's the name of the company?"

"Sleuth Software. The software package is called 'Sherlock.'"

"And you have a copy of Sherlock here?" Phil was staring at the multitude of windows on the computer screen.

"Yes, and he probably is dying to meet you both." Allen pressed a button turning on the computer's speakers. He leaned forward. "Sherlock?"

A deep voice boomed from the computer speakers. "Yes, Allen?"

Phil and Ann jerked back in surprise.

"I'd like you to meet Ann Stevens and Phil Conley."

A message flashed on the status window.

09:20 AM TWO UNKNOWN SUBJECTS IDENTIFIED AS ANN STEVENS AND PHIL CONLEE.

Allen corrected Phil's last name in the Voice Recognition window as Sherlock replied with a series of canned phrases. "A pleasure to meet you both. Who is seated in the interrogation chair?"

Ann jumped up instinctively. "The what?"

"Ann is," Allen chuckled. "I thought you wanted to test the system?"

"Err... I do, I guess." Ann sat back down uneasily, staring at the metal boxes on the table. Phil pulled a chair next to Allen and sat down staring at the computer screen.

"Fine. Let's get started, then. Sherlock?"

"Yes, Allen?"

"Open a data file."

"What is the nature of the file?"

"Interview of subject, Ann Stevens."

"File opened."

"First, we'll start with some easy background questions to establish a reference point."

Ann nodded as Allen turned the computer speaker off and picked up a clipboard with a questionnaire on it.

"Ann, what is your full name?"

"Ann Marie Stevens."

Several indicators jumped in the one of the windows. Phil saw a number flash in a small window labeled 'advice'.

"Allen, what did you say the number in the 'advice' window means? It read 98 percent for a few seconds."

Allen glanced up from the clipboard. "That's the percent probability that Ann is telling the truth."

"Do you want me to tell a lie?"

"Yes, but I would rather ask you several questions and then let Sherlock tell me which answers were lies."

"Okay."

The infrared images on the screen started moving and a pattern appeared in the Voice Stress Analysis window each time Allen asked questions and Ann answered. Phil watched Allen write the 'advice' value on a questionnaire form after each response.

"How old are you?"

"Thirty-three years young."

"Ann, just answer the question."

Ann laughed. "All right."

"How long have you been married?"

"Eight years."

"How many children do you have?"

"Two."

"What are their ages?"

"Six and four."

"How long have you lived in your present house?"

"Three years."

"What kind of a car do you have?"

"A Mercedes."

"How many years have you owned a medical equipment business?"

"Nine."

"One more... how long have you known Phil and me?"

"Since our Freshman year in high school."

Allen put the clipboard down and corrected a few misspelled words in the dialogue captured in the 'Voice Recognition' window and typed "Run Analysis." A few windows closed. Allen turned the computer speakers on and sat back in his chair.

"Sherlock?"

"Yes, Allen."

"Did Ann tell a lie?"

"Yes."

"Which answer was incorrect?"

"The response to the sixth question was not truthful."

Allen could tell that Phil and Ann were impressed.

"You're right. We've lived there four years."

"Sherlock, was Ann's last response true?"

"Yes."

"Sherlock can evaluate all statements, as well as answers to questions."

Phil was still skeptical. "But these were easy questions, and nothing she's particularly interested in keeping a secret. Why don't you ask her something hard like... has she ever cheated on her taxes?"

Ann suddenly sat upright. "Hey!!! Let me ask Phil some questions."

Allen laughed. "Guys, no need to get excited. This is just a demo, remember?"

Ann and Phil looked at each other. "Ann, are you thinking what I'm thinking?"

Ann was leaning forward in her chair, suddenly excited. "I sure am! Allen, who else knows about this system?"

Allen was surprised at their sudden interest. "I've only tested it on my technician, and the night janitor who thought it was a lot of fun."

"The last response of Allen Atkins was not true," said Sherlock.

That caught Allen by surprise and he blushed.

"Err... all right, I also tested it on Janet Turner. I just wanted to make sure it was working before I told you guys about it."

Ann wagged her finger at Allen when Sherlock commented, "The last response of Allen Atkins was not true."

Allen's face was a red as a beet. "All right, I would like to date her, and I wanted to see if she was dating anyone." Before Sherlock could comment again, Allen turned the computer speakers off. "Why do you want to know?"

"Phil and I think you shouldn't tell anyone else just yet."

"Why? It's certainly not a commercial product at this point."

Ann was thinking hard about her upcoming expansion. "Allen, could you let me borrow this equipment for a few days?"

"What! You know this isn't my stuff. It all belongs to NASA... except for Sherlock. It wasn't being used right now, so I checked it out and started putting this system together on my own time."

"What if it were in the interest of gathering more data to test the system?"

Allen knew Ann was always willing to try new things, but she was sometimes too impulsive. "What do you want to do?"

"I need to hire a few more employees, and I sure could use it to see if they are lying during the employment interviews."

Allen laughed loudly. After a moment Phil laughed too.

"You can't do that! Phil, didn't you say last night there are laws against using lie detectors to screen prospective employees?"

Phil nodded. "It's only permitted in a few selective cases, like national security for example."

"And what happens if they catch you using it in violation of the federal law?"

"You can be fined and even put in jail."

"See, Ann. I don't want to be a party to something illegal. I could lose my job for sure, and I like my job!"

"But what if the interviewee didn't know it? It doesn't produce any written records, does it?"

"Uh... only when you tell it to. But how could the interviewees, as you call them, not know what is going on, with all this equipment around?"

"We could just say we're recording the interview so other managers can see it later. Or some story to that effect. This could be a great help in determining if your system works."

Allen was tempted. He really needed more data in the database for the Artificial Intelligence software to function correctly. "I don't know. Somehow it doesn't seem right."

"Why not? If this system is as good as you say it is, then innocent people wouldn't have anything to worry about. As I understand it, the main objection to lie detectors is that they label innocent people as liars, more often than they clear people who are liars. Or something like that."

"But I can't just take this stuff out the gate. They do random searches of cars, and if they caught me with it, I would be fired for sure."

"Allen, what if Ann and I were to buy a complete set of this hardware? Would you set it up?"

"Well, sure. I guess that would be okay, since I wouldn't be using NASA's stuff."

Phil looked around the lab. "What would the hardware involved cost?"

"I don't know, but I would guess it would be less than $100,000."

Ann coughed while Phil's eyes grew wider.

"Did you say less than $100,000?"

"Yes. Is that more than you thought?"

"A little. This system would have to be pretty good, for anyone to pay that much money for it."

"But there's a lot of stuff here: a miniaturized spectrophotometer, a carbon dioxide laser, a computer, an infrared video scanner, a voice stress analyzer, voice recognition software, a voice synthesizer, an Artificial Intelligence software development package..."

"Okay, okay!" interrupted Ann. "But it's still a lot of money. Phil, what would a nearly foolproof lie detector be worth to the police?"

Phil thought about that for a moment. "If it were truly almost foolproof, it would be invaluable." He suddenly remembered a friend whose son was on trial for murder.

"You know, I might know someone who would pay anything if it could get his son out of jail." Phil took out his cell phone, but there was no signal strength. "There's no signal in here. Where is the nearest wired phone?"

Allen frowned. "I don't think it's ready for that, but there's a phone's in the main lab, on the table near the main laboratory door."

Phil looked at his watch. "It's only 7:30 in the morning in Los Angeles. He's probably still at home. I'll call him."

As Phil left to make the call, Ann walked over to look at the computer screen. Several windows were open, and a message was flashing "No infrared data."

Allen watched her for a moment. "Want to see if you can fool it?"

Ann laughed. "No, thanks. You know, Allen, you've come up with some pretty crazy things over the years, but this one might just be the mother lode."

Michael Stone yawned and stretched his aching muscles. He glanced at his watch. 7:30 AM. «Thirty lousy minutes to go,» he muttered as he put his headphones back on and settled back in his chair. At six foot seven, his long legs barely fit under a makeshift desk cluttered with food wrappers and empty coffee cups. The converted cargo van was so crowded with sophisticated listening gear, he often complained to his boss that he barely had room to scratch himself. It had rained all night, and the rain's steady drumbeat on the roof had made it even harder to stay awake.

La Brea Security had been hired to eavesdrop on the mansion of Adam Daniels. Two men identifying themselves as Security and Exchange Commission officials had presented a judge's authorization for a wiretap, and the president of La Brea, Josh Williams, had been more than happy to oblige. Daniels was an extremely wealthy investor in junk bonds and other high-risk corporate securities, and it seemed reasonable to Williams that Daniels could have run afoul of the SEC watchdogs.

Stone had rotated into the dreaded night shift, and none of the half-dozen cell phone lines his was monitoring had been used even once. What boring people! He absent-mindedly scratched his scraggly beard, snapping to attention only when Daniels' cell phone rang. "The main guy... oh, yeah!" he said softly. "Let's go, baby. Spill your guts."

Williams would be delighted to hear about this one.

Ann rubbed her chin thoughtfully. «Do you think some of this could be adapted for a respiratory monitor for trauma patients?»

Allen was skeptical. "It's certainly possible, but it would only require a part of this system. And NASA would have the rights to it, so you would have to work something out with them."

"But then they would know that you've been talking to me about it!"

"Oh, yeah! Come to think of it though, the only thing proprietary is the miniaturized spectrophotometer. It had to be made as small as possible to minimize the size and weight during a space flight." He mentally constructed a possible system. "A regular detector would work though. Of course, it would be a lot larger and heavier."

"And more expensive!"

"Actually not. This one is a prototype and it cost a fortune to develop."

"So, a regular detector would be less? Would that mean a whole system would be significantly less than $100,000?"

"I could get the price on everything and tell you exactly if that is what you want."

"Yes, please. If I know that, I may be able to find a market for it."

Allen was writing a note to himself as Phil returned.

"Allen, could you have a second system ready to put to a real-world test in about a month in Los Angeles? I told my friend about this, and he said bring one out as soon as possible. The cost doesn't matter. How about that?"

Ann laughed. "Sounds like you have made one sale already."

"A friend of mine, Adam Daniels, is loaded and his son Ross is in jail accused of murdering someone at a party. His son claims he was sleeping when it happened, but there are some witnesses who claim they saw him leave the room where the body was found."

"If the witnesses truly believe it's him, then a lie detector won't help."

"We think they may be mistaken or covering up for someone else."

Allen seemed unsure. "I don't know... what's the timing on the trial?"

"It's set to start in about two months. How long would it take to put another system together?"

"That would depend on two things: the delivery time on the equipment you see here, and some time to add additional data to the system to increase its knowledge reference base."

"I can get you a cash advance to buy everything immediately. How can we get the data you need?"

Ann interrupted. "Phil, couldn't you figure out a way we could use this system to interview some potential employees for me? We could gather a lot of data for Allen's database. We have over a hundred applicants for a few openings."

"I don't know... You would have to tell them what you're doing, and they would have to voluntarily agree to it. You couldn't make it a requirement."

"That's okay. Even if half of them agreed, we could still get a lot of data."

"That would help get the range of normal responses when they answer the standard questions like their name, age and address… yes, that would do it."

"So we can be ready to go to LA in about a month?"

"I have a job here, you know."

"Look, I know you love your job, but this is a golden opportunity to test your idea. Could you take a leave of absence?"

"Well, I do have four weeks of vacation left over from last year.…"

"Great! How about an all-expense-paid vacation to Los Angeles in about a month?"

"Somehow, I don't see a lot of free time for sightseeing in that question."

"If this works out, you could be a very wealthy man. You wouldn't even need to work anymore. Believe me, this guy knows how to repay a debt."

"Hey! What about me?" exclaimed Ann. "If I can help gather the background data needed, would he be grateful to me too?"

Allen and Phil laughed.

"I'm sure if there is a monetary angle to be found in all this, you will. So… how about it, Allen?"

"Say yes, Allen. I haven't been to LA in a while," urged Ann.

"What about your kids?"

"My mom can take care of them until I get back. And… I would go home every weekend to make sure they're okay."

Allen sat down and stared at the computer monitor for a moment.

"Okay. Let's see if this thing can flush out some liars."

Ann hugged Phil as he shook Allen's hand.

The phone call on Daniel's personal cell phone had been related to his son's upcoming trial for murder and not to financial matters, but Josh Williams had been ordered to report anything they found immediately to a special contact's phone number. He clicked open an electronic Rolodex app on his phone and zipped through the names until he spotted the contact's number. He punched the number in his cell phone.

"Yes?" replied the contact.

"Uh, hello this is Josh Williams with La Brea Security. I was instructed to pass on everything we found concerning the Adam Daniels matter?"

"Oh, yes, go on," urged the contact.

Jake «the Snake» LoBlanco was the closest thing left to what might be termed a mobster in Las Vegas and his influence was felt in almost every illegal activity. His moniker was due to an unfortunate skin condition that flared up from time to time and he dealt swift and sure punishment to those that used that nickname in his presence. Rumor held that he had personally dispatched the last unfortunate soul to do so.

LoBlanco stretched, put on a silk, monogrammed bathrobe, and walked out onto the balcony of his lavish penthouse. He surveyed the morning traffic and the clouds gathering over the distant mountains. The air was brisk, and he tightened his bathrobe as his housekeeper placed his morning coffee and newspaper on a table and stood for a moment. Lucia Hernandez had been his housekeeper for over three years, a virtual record among his employees. She had a deep respect for his apparent command over his "troops", as he called them, that was mixed with an equally strong fear of displeasing him. She hoped he wouldn't mind a visitor this early.

LoBlanco sat down and picked up his coffee cup and saw her waiting to speak to him.

"Yes, Lucia?"

"Mr. LoBlanco, Brian Limpanatti is here to see you. Do you want me to tell him to come back later?"

"No, I'll see him now."

Lucia left and a moment later, Loblanco's top lieutenant appeared and waited nervously for LoBlanco to acknowledge him. LoBlanco glanced up and motioned Limpanatti to sit. He finished his coffee.

"What's going on, Brian?"

"I just got a call from Williams at La Brea. Daniels got a call this morning from an old friend offering to help with his son's case. We did some checking, and the guy is an assistant DA in Houston."

LoBlanco frowned. "Daniels can afford an army of lawyers. What's the deal? Why did this old friend wait all these months to offer his help?"

"Williams is faxing me a transcript of the call. All I know, is this guy mentioned some new lie detector that he said could help prove his son's innocence."

"Lie detector? Do they think the witnesses are lying?"

"I don't know, boss."

LoBlanco stood up, walked to the balcony railing, and stared off into the distance. After a moment he looked back at Limpanatti. "I don't like this."

Limpanatti felt a nervous twitch in his stomach. "What do you want me to do?"

"First, call Ryan and see if there have been any leaks, and if there have been, plug them."

"Yes, boss."

"Then find out what you can about this new lie detector."

"Yes, boss."

"Oh... and, Brian?"

"Yes, boss?"

"Do this quietly. Get some help if you need it."

Limpanatti grinned. "Yes, sir!" He turned and walked quickly through the penthouse, relieved that LoBlanco didn't blow up at the news, and happy that he had an opportunity to show him what he could do. He was whistling to himself as he waited for the penthouse's private elevator.

CHAPTER 4

Janet pushed open the door to Allen's lab and entered looking for him. "Allen!"

The lab appeared to be empty, and she walked to the small lab in the back, but it was deserted also. She saw a small white pad of paper next to the computer and sat down to leave him a note.

Dear Allen:
My parents just called, and they can't make it this weekend. If your offer is still good for dinner and a movie on Saturday, give me a call at...

She started to hum to herself as she wrote and didn›t notice the microphone on the table in front of her. The computer screen jumped to life as several windows popped open. Several observations from Sherlock appeared in the status window.

02:30 PM LAB ACTIVITY FILE OPENED. ONE UNKNOWN PERSON IS IN RANGE.

Janet glanced up at the screen but didn't pay it very much attention. She continued writing until the computer BEEPED at her twice. Several more windows opened on the screen and their status changed from Inactive to Active.

A box on the screen labeled "advice" had a red warning label that blinked "Insufficient data for advice." Her eyes were drawn to another blinking warning sign:

02:31 PM NO ACTIVE INFRARED IMAGING DATA, NO ACTIVE CO2 DATA.

She also saw a blank line in the "Voice Stress Analysis" box. She was trying to figure out what this all meant when Ron Johnson, Phil's technician, sauntered in.

02:32 PM A SECOND UNKNOWN PERSON IS IN RANGE.

Ron was in his late 30's and divorced. He must have tried dating every available female at NASA and had even asked Janet out a few times. Even though he was a little overweight and slightly balding, he somehow pictured himself a ladies' man. He was wearing his usual white lab coat with his employee badge pinned upside down. He smiled when he saw Janet.

"Hi, Janet? How are you doing?"

Oh, oh, she thought. *I better make this quick.*

02:33 PM SECOND PERSON IDENTIFIED VIA VOICEPRINT AS RON JOHNSON.

"Hi, Ron. I'm doing fine. I wanted to see Allen, but he doesn't seem to be around, so I'm just writing him a note."

02:33 PM FIRST PERSON IDENTIFIED VIA VOICEPRINT AS JANET TURNER.

"Oh." He sounded a little disappointed. "He was here a little while ago. He must have gone to a meeting or something."

Ron forgot about Allen's experimental setup and sat down on the "subject" chair across from her.

"So, how are your finals going?"

Janet was secretly pleased that Ron remembered she was studying for a master's degree. From her experience, most single men weren't good listeners and usually spent most of a conversation talking about things they were interested in. In his gung-ho dating mode, Ron was a good listener and always eager for conversations with potential dates. It was too bad Ron wasn't her type. Allen probably wouldn't have remembered her finals were this week.

"Great! My last final was last night."

The red warnings on the pop-up windows changed to green and the computer beeped at Janet again, drawing her attention to the screen. Several messages appeared in the Status window.

02:34 PM INFRARED IMAGING LINK RE-ESTABLISHED.

02:34 PM CO2 ANALYSIS LINK RE-ESTABLISHED.

02:34 PM VOICE STRESS ANALYSIS ACTIVATED.
02:35 PM AI ALGORITHM ACTIVATED.
02:35 PM ADVICE CALCULATIONS BEGINNING.

The advice numerals were blinking red. Ron picked up on Janet's confused expression.

"What's wrong?"

Janet's glance alternated between Ron and the screen. "Oh, nothing," she replied wistfully.

Ron had often boasted to her of his latest flings in hopes of changing her apparent lack of interest in him. "Hey, did I mention I'm going out with Rhonda now?"

Janet didn't immediately answer as her attention was drawn back to the screen by another beep. The Advice numerals were green, and the value was 90%. Janet frowned. What did he say? When she looked back at Ron, he was cleaning his fingernails with a pocketknife. Did he say Rhonda?

"Rhonda Morrison? I thought you were going out with Wendy Gonzalez?"

Ron was strutting a little as he replied. "Oh, I had to break that off. She was starting to get too serious. She probably would be wanting a ring by now."

A window blinked on the computer screen, drawing Janet's attention. The Advice number was reading 20%, and had changed from green to red. A growing realization of the system Allen had apparently put together dawned on Janet as she looked back at Ron. She tried not to laugh. "Really, Ron? I heard she dumped you for some project manager."

Ron was more than a little surprised Janet seemed to know this. He immediately became defensive. "That's not true! She's probably just saying that, so people won't think I dumped HER."

The computer screen blinked again, and this time the Advice numerals were red and the value was 10%. It was hard for Janet to keep from laughing out loud.

She held her hands up and gestured. "That sounds like an 'I caught a fish this big, but it got away' tale."

Ron seemed uncomfortable, eager to leave. "Well, you can believe what you like. I have to get back to work."

As Ron stood up to leave, the Advice numerals flashed a red value of 15%. Janet laughed and Ron frowned at her as he stood up.

"I'll tell Allen you came by. See you later."

He left, and Janet sat staring at the screen for a moment. She suddenly remembered her earlier conversation with Allen when she was sitting in that chair and became angry. She picked her note up and stuffed it into a pocket of her lab coat as she stood up and stalked out of the lab.

Jim Hall and Steve Jenkins were chatting with Wendy Gonzalez, the department secretary for the group Allen worked in, as Janet walked up to them. Jim was about Allen's age. Slender, polished, and well dressed, he probably could have posed for a gentlemen's quarterly type of magazine. Steve Jenkins was older, distinguished looking with a little gray in his temples. Jim saw Janet first.

"Hi Janet! We were just talking about our plans for the weekend. Steve is going offshore fishing and I'm going tubing on the Guadeloupe River. How does your weekend look?"

"Well, for a while I thought I'd be going to dinner and a movie, but now I'm not so sure."

"How come?"

"I think my plans have just changed. Wendy, do you know where Allen is?"

Jim looked at his watch. "Oh, we're late for a project meeting. We'll see you later, Wendy." He said as Steve picked up his briefcase from the floor. They walked off talking as Wendy looked up from her computer.

"Allen stopped by a while ago to say that he will be gone the rest of the day. He also said he is going on vacation next week."

"He's going on vacation? He didn't say anything about that!"

"Oh? Are you two going together now?"

Janet blushed. She forgot that Wendy was the biggest gossip she had ever known. She would have to be more careful in the future. "He asked me out this weekend, but I thought my parents were coming in, so I had to say no. They just called to say they couldn't make it, and I was trying to find Allen to let him know." She paused, gritting her teeth. "I also want to ask him about something."

"Why don't you call him at home? I have his number if you want it."

"Yes, please."

Wendy texted the number to Janet's phone, then looked at her watch. "Oh! Time to go! I have a hot date tonight... I hope you can get in touch with him."

As Wendy started putting some things in her purse, she saw Janet staring at the number in the text, apparently lost in thought. She smiled as she thought of some people that would be interested to know Janet was dating Allen.

Allen, Ann, and Phil were in Allen's apartment engrossed in the assembly of the second prototype system, unpacking equipment from boxes, connecting cables and testing various pieces under Allen's direction. Phil looked at an unusually small laptop computer with a metal case. He had never seen one quite like it and didn't recognize the manufacturer. He looked at Allen.

"Are you sure this laptop is powerful enough for the system? You had a really powerful desktop computer in your lab."

Allan laughed. "Don't let the size fool you. That's one of the most powerful desktops made. It has dual CPUs, 1000 megabytes of memory and...."

"Okay, okay! I was just asking."

The phone rang, and Allen hurried to pick it up.

"Hello?"

There was a pause and Janet began in a loud, angry tone. "I just wanted to let you know my parents can't make it this weekend."

Allen wondered why she was so angry. "That's great... I think. What's wrong, Janet?"

"We need to talk about something."

Allen glanced up at Ann and Phil who were continuing to put the equipment together, apparently oblivious to the phone conversation. Allen looked at his watch.

"Sure. How about meeting me for dinner and we can talk about whatever you want?"

Janet didn't reply immediately, but when she did her tone seemed less angry. "Fine. Where?"

"How about the new restaurant near the mall? I can pick you up in fifteen minutes."

"No, I'll meet you there."

Allen paused, wondering what the problem was. "All right. I'll see you in about fifteen minutes."

"Okay."

Allen heard her grunt angrily and disconnect the call. He stared at his phone for a few seconds before putting it in his pocket. "I wonder what's going on?" he said to himself.

Ann and Phil were still going strong with the assembly. "Hey, guys... can you work on that for a while without me? I think Janet needs to talk about something really badly."

"Sure, go ahead. I really need to go anyway. Dan is flying home tonight. I'm still hopeful we can work some things out," replied Ann.

Phil looked at his watch. "Oh, my God! I need to go too. I have a date with Joanne in 30 minutes. I'll see you guys first thing tomorrow morning."

Ann and Phil left as Allen went into the bedroom to change clothes. He paused to look at himself in a mirror. "I wonder what's going on?"

Allen walked into the upscale Mexican restaurant and was approached by the hostess.

"One, Señor?"

"I'm supposed to meet someone here. Woman, about this tall, dark hair, beautiful smile." Allen held his hand up about shoulder high to the hostess.

The hostess smiled. "Are you Allen?"

"Yes?"

"Right this way." She led him to a table in the back where Janet was drinking a Margarita. Allen sensed she was upset as he sat down opposite her.

"Hi!"

She glared at him for a second before replying. "Hi."

The hostess was still standing next to Allen. He looked at her. "I'll have a Margarita also."

"Okay. The waitress will be right over."

The hostess left, and Allen and Janet stared at each other for a moment before Janet began. "I came by your lab today to tell you my parents can't make it this weekend."

Allen could tell from the tone of her voice that she was about to unload on him. "That's great... so why are you so angry?"

"I happened to be in your little cubby hole, writing you a note, when Ron came in. We were talking about his latest romantic conquests when your computer started beeping at me."

Allen swallowed hard. "It did?"

"I'm not an expert in certain areas, so it took me a while to figure out that you have built some sort of a lie detector system... or something like that... right?"

Allen was feeling guilty at using the system on her and even more so now that she had caught him. "Yes," he replied quietly, avoiding her stare.

"And you used it on me this morning when you asked me out, didn't you?"

Allen felt his face flush at having been caught. "But I didn't do it to hurt you or be mean or anything like that."

"Then, why did you do it?"

The waitress walked up and put a Margarita in front of Allen. She handed both a menu.

"Hi! My name's Juana. I'll be your waitress tonight. We have a 2-for-1 special on fajitas tonight. I'll be back in a few minutes to take your order."

She left as Allen put one of his hands on one of Janet's hands. She didn't take it away.

"Janet, I've been wanting to ask you out for a long time, but I didn't know if you were going out with anyone, or in a relationship or something. I guess I thought I could bring the conversation around to that and find out for sure without you knowing."

"Why didn't you just ask me?"

"I've been out of the market, so to speak, for a long time. I was married to Mary for almost ten years. I guess I'm not used to being single again and asking things like that straight out. I know you do, and I wish I could be like that too."

Janet seemed to ponder this. "Well... what did you learn from that machine?"

"That everything you said was true, and you didn't say no when I asked you out. I took that as a good sign."

Janet smiled and puts her other hand on Allen's hands. "For someone as smart as you obviously are, you have a lot to learn about women... or some women... or me in particular."

"I would like to learn a whole lot more about you."

"I might like that as well. But first, I think we should agree that there won't be a need for a lie detector from now on to find out what the other thinks."

Allen held up his right hand, "I swear to tell the truth, the whole truth, and nothing but the truth."

Janet laughed.

CHAPTER 5

Phil held the umbrella tightly as he ran through the parking lot in the rain. He glanced at his watch as he hurried up the stairs to Joanne's apartment. Although it was only their third date, it could turn out to be a very special night. Phil had dated numerous women and even had been engaged once, but a tough work schedule and a passion for his work had been hard on his relationships. Many had started well only to falter in a fairly short time. He was determined this one would be different. He cleared his throat, straightened his tie, and knocked on Joanne's door.

Joanne also had had many relationships that hadn›t worked out and had been engaged twice. She had even been left standing at the altar when her fiancé had gotten cold feet, and then had the nerve to call her from the airport and tell her he was leaving town. Her chance meeting with Phil had excited a passion she hadn't felt since high school. It was much more than a mature appraisal of Phil as a potential lover. Although she didn't believe in love at first sight, she had been strongly attracted to him from the very first moment they were alone. When she described the feeling to her best friend, she had called it "chemistry."

She had passed the physical-attraction aspect immediately. She had guessed that Phil was only a few years older than her 30 years. He was athletically fit, and she had an affinity to blond haired men that she hadn't even realized until she thought back on her previous boy friends. Phil was also a gentleman in the classic sense, opening doors for her, holding a chair for her in a restaurant, generally making her feel special. Joanne didn't know Phil had been attracted to her in the same manner. Phil didn't even realize he was enamored of red-haired women.

Their second date was special for both as they dined at an elegant restaurant and went to a theatrical play. When he kissed her goodnight, it took all of her strength to keep from asking him to stay with her. She was afraid that premature intimacy, based only on a physical attraction, would lead to the usual breakup in a few months. This pattern repeated itself in her life a few times, and she was determined it wouldn't happen again.

For their third date, she offered to cook dinner and he was bringing a list of UserIDs and passwords for all his streaming services so they could spend a quiet evening alone. She was fascinated with his stories of criminal cases and really wanted to get to know more about him.

Joanne answered the door and Phil held out a bouquet of flowers. She wondered why Phil was wearing a suit for a quiet evening alone as she took the flowers, and stepped back to let him in. As he passed her, he put an arm around her and kissed her.

"I would have been here sooner, but I didn't have time to change. Something sure tastes good. What are we having for dinner?"

Something about Phil seemed to raise her blood pressure and she knew it wasn't his cologne. She had to take a deep breath to slow her heart down.

"Pot roast with mashed potatoes and carrots, and corn-on-the-cob, with pecan pie for dessert."

Phil stared at her for a moment. "Are you serious?"

"Yes, why?"

"I don't think I've had that since I left my parents' home."

Joanne smiled as she took his overcoat and umbrella.

Phil helped Joanne clean up the kitchen and they settled into her sofa to watch some streaming movies. Although they began by sitting closely together, they somehow had managed to wind up lying on the sofa. As they snuggled closer, Joanne couldn't resist him any longer.

"Phil?"

"Yes?" he replied between kisses.

"If I said there is a difference between sex and intimacy, would you know what I mean?"

"I think I know what you mean."

"Well... we haven't known each other very long, but I am attracted to you and I..."

He put a finger over her lips.

"I've felt the same about you from the first moment."

Her eyes must have given her away. Phil smiled. "And... I'm not afraid of commitments."

She pulled him even closer and whispered in his ear. They stood up and kissed for a moment and she took his hand and led him into the bedroom. The curtains were open, and the room was dimly lit from a light in the parking lot. Joanne started to close them, but he quickly stood behind her, taking her hands in his and wrapping his arms around her.

"I want to see you," he whispered and kissed her neck.

As they looked out into the night, the rain ended and a full moon peeked out of the clouds. He kissed her neck again and let go of her. Moonlight spilled over her hair and shoulders as she turned to him. He sat down on the bed to watch her undress. Her heart was pounding and she had somehow managed to take her jeans off but her hands fumbled with the tiny buttons on her blouse.

"Let me," he whispered.

As she stood in front of him, she felt his hands on the back of her legs pull her closer to him and then slide up her sides and onto her breasts. She put her hands on his shoulders and felt him unbutton her blouse. She let it fall to the floor and took her bra off. He stood up and took his shirt and pants off. They hugged each other for a moment content to feel each other's body. Her head was on his chest, and she could hear his heat beating. She felt his hands push her panties down and she could tell that he had taken his shorts off.

"Phil?" she whispered.

"Yes."

"It's been a while for me."

He knew what she meant. "I won't hurt you."

She pulled the covers back and laid down. He knelt at her feet, and started kissing her toes. She took a deep breath when she felt a hot tongue on her ankle that started sliding slowly up her leg. Joanne lost track of time as Phil's tongue seemed to find every erogenous area of her body. She couldn't even remember how many times her passion had reached a crescendo. When they finally made love, it was almost anti-climatic for her, but not for Phil. He seemed determined to release pent-up desires, until she heard him laugh.

"Why are you laughing?" she whispered in his ear.

"You better tone it down or the neighbors will think you're in trouble."

Joanne didn't understand immediately, then it dawned on her. She hugged him as she whispered in his ear. "See what you do to me!"

CHAPTER 6

Mary Atkins was settling in for another lonely evening, watching television. She had received the house as part of her property settlement with Allen, but she had to sell it and most of the furniture to pay her part of the legal bills. Her new apartment was pretty barren. She had begun munching on a bag of microwaved popcorn when there was a knock on the door. When she opened it, she was surprised to see Gene Abrams leaning against the doorframe, obviously drunk. She could see his motorcycle in the parking lot. Gene usually didn›t dress the part, but tonight he was wearing a denim jacket and leather jeans. His hair was even longer than the last time she had seen him, and his normally scraggly beard was now over six inches long. She almost didn›t recognize him.

"Gene! I thought you were going to be in LA for a few more months?"

"Things didn't work out as I expected, so I came back early," he said unevenly.

"Well, come in, then."

He motioned to his motorcycle. "No. Let's go get a drink and talk."

Gene could hardly stand up and she didn't relish the idea of riding with him. "Okay, but you look tired. I'll drive. Just a second."

He waited while she picked up her purse from the sofa and locked the front door. Mary had to stifle a laugh when Gene stumbled and banged his head getting into her car. After a short ride, they entered the "Rowdy Times", the crowded and noisy nightclub where they had first met. With long unkempt blond hair, a fairly muscular build and several prominent tattoos, Gene wasn't exactly Mary's type, but a few months after her separation from Allen, she was beginning to feel out of place with her married friends. One night, she had been so depressed and lonely she had asked Gene to drive her home and spend the night. Gene told her later that he was surprised a "classy" lady like her would even go out with a motorcycle mechanic. Shortly after they started dating, Gene left for Los Angeles to help an old friend get a new motorcycle shop going and to try and find enough money to start a dealership of his own.

Gene led her through the crowd and managed to find two empty stools at the bar. He ordered a beer while Mary ordered a harder drink.

"So what's going on, Gene? Why did you come back so soon?"

"The deal I was trying to make went sour, and I had to beat it out of there."

The drinks arrived, and Gene almost finished the beer in a single drink. Mary frowned. "Are you all right?"

"Yeah... did you ever get the rest of the settlement money?"

"Jason said it should be any day now... why?"

"Do you have any money you could lay your hands on in a hurry?"

"The only money I have, is a savings account that I've been using to help pay the bills. There's about ten or eleven thousand left, I think. You could have some of that."

"That's not enough. I need at least thirty thousand."

"Thirty thousand! Whatever for?"

"I have some gambling debts I have to pay on or I'm going to be a chalk line on the evening news some night."

Mary stared at him in amazement. "You mean they would kill you?"

"If I don't pay something soon, they'll write that debt off and me along with it."

Just as he finished his beer, a HUGE burly man next to Gene picked up a tray of drinks to take to his table. The club was crowded, and he accidentally bumped Gene as he passed. He looked back and commented with a deep, booming voice. "Oh... I'm sorry." He started to walk off, but Gene pushed him, almost knocking the drinks off the tray.

"Watch it, asshole."

The huge man stopped and glanced back. "I said I was sorry. Go screw yourself!"

Gene stood up. "You stupid jerk!"

The man stopped. He put the tray with drinks on a nearby table and faced Gene. "What did you say?"

Mary quickly pulled on Gene's sleeve. "Gene, don't."

Gene jerked his sleeve away. "I said watch where you're going, asshole!"

The man pushed his chest right up to Gene's face. "No, you didn't. You called me stupid. I may just mop the floor with this nice pretty hair of yours."

Mary tugged on Gene's sleeve again. "Gene, let's GO!"

Gene pushed her back, staring up at the huge man. "I'd like to see you try it."

A few people began crowding around them to watch. Mary picked up her purse and moved back to a wall out of the way. The huge man quickly wrapped a big burly arm around Gene's head and hit him on the head with a fist. Gene dropped to the floor, dazed for a moment. He leaped to his feet and punched the huge man in the face. It knocked him back but not down. He rubbed his face and growled. The onlookers backed up some. The commotion sent people in nearby tables scurrying out of the way. The bartender reached under the bar and held a shotgun out of sight. He shouted to Gene and the huge man.

"Hey! No fighting in here. I'll call the cops. Take it outside."

Gene saw two men dressed in black enter the nightclub and begin talking to a waitress. He looked desperately around until he spotted Mary. He glanced quickly back at the huge man. "We'll dance later, asshole. I gotta go." He walked quickly over to Mary and grabbed her by the arm. "We gotta get outta here. They're here."

She stared blankly at him. "Who's here?"

"Never mind. Come on!" He pulled her to the kitchen door, and they left through the back delivery entrance. Some onlookers patted the huge man on the back. He picked up his drink tray and walked to his table. The two men dressed in black walked slowly around the dance floor looking for someone.

Mary was worried and drove home on autopilot. Gene was resting his head back on the seat rest. His eyes were closed.

"What are you going to do?"

Gene didn't open his eyes. "I need to find a job. I know a loan officer at a bank. He owes me a favor and said he could get me a loan, but first I have to have a job."

"I know a company that may be looking for a truck driver."

Gene opened his eyes and sat up looking at Mary. "I used to drive a truck. Do you know someone there I could talk to?"

"I know everyone, even the owner of the company. Allen and the owner started the business."

"What! You mean Allen worked somewhere else than NASA?"

"It was a long time ago, when Allen and his friend Ann Burger were in college. Allen was always coming up with crazy inventions. One time he came up with an idea for some new medical gadget. He even got a patent on it. Allen didn't have any money, but Ann's father was a partner in a law firm in Houston and loaned them the money to start the business."

She briefly reminisced about the past, before she continued.

"Allen is a big thinker but a poor businessman. Fortunately, Ann excelled in business. They started pretty slowly, but Allen was always wanting to tinker with some new invention. Eventually, he sold out his half to Ann and used the money to go to graduate school and make some new 'toys' as he calls them. None of them ever panned out, of course."

Gene seemed unusually interested. "How about the business?"

"It's doing pretty well. I think they have over a hundred employees now. Allen could be pretty wealthy now if he'd just stuck with the company."

Gene looked out the window, then back at Mary. "Who can I call?"

"Probably the head of personnel, Margaret Simpson. I'll look up her number when we get to the apartment."

"If I can't find something soon, I may have to go into hiding and hope they won't find me."

"Damn, Gene. How much do you owe them?"

Gene laughed. "More than you got for your house."

Mary stared at him in disbelief.

The «Rowdy Times» bar was not quite as crowded as earlier in the evening, but it was still noisy as Allen entered with two fellow researchers, Jim Stall and Steve Rawlins. Jim and Steve were wearing suits from a project presentation but had taken their coats and ties off. Allen looked as if he has been playing golf. As they entered, they passed the two men dressed in black who had entered earlier, looking for Gene. Allen, Jim and Steve sat down at a table near the bar. As the waitress took their orders, Allen was amazed to see Janet at a table with a girl friend. He observed two "cowboys" trying to pick them up and the fact that Janet and her friend obviously didn't like it.

Allen glanced briefly at his friends. "Excuse me a minute, guys. Janet Turner is here with a friend. I wouldn't have thought she'd come to a place like this.

Jim laughed. "You mean, to associate with low-lifes like us?"

"I didn't mean it like that. She likes classical music."

"Oh, go ahead. You probably wouldn't be much company right now anyway."

Allen grinned at them and casually sauntered over to Janet's table. He positioned himself next to one of the cowboys who were still trying to pick up Janet and her friend. Janet's friend appeared agitated. "Buzz off, cowboy. We're just here to listen to some music and have a few cool ones."

The taller cowboy moved even closer to them. "Oh, come on, honey, just one dance with Dave and me. You won't be sorry."

"We're not interested!" exclaimed Janet.

Dave was insistent. "Well, we aren't going to leave until we get a dance, are we, Jack?"

Jack laughed. "Hell, no. Come on now, just one dance. You'll like it...."

Allen interrupted. "Hi, Janet! I didn't know you liked Country music."

Janet was so surprised to see him. "Allen!"

Jack was now annoyed. "Hey! Beat it, Mac. We were here first."

"It doesn't sound like they're interested."

"Well, that's none of your damn business, now, is it?"

His friend pushed Allen a little. "Get lost."

Jim and Steve were watching, and Jim started to get up.

"Jim, don't bother. Allen can handle this."

"But there are two of them!"

Steve laughed. "It wouldn't matter if there were ten of them, Allen can still handle it."

"Huh?"

"Don't you know? Allen earned a third-degree black belt in Karate while he was in high school."

"What? How come he never talks about it?"

"He was in some Junior Nationals tournament or something when he hurt someone. It was an accident, but Allen felt really bad about it and sort of dropped out of martial arts. Look... just watch and learn."

They watched as the two cowboys moved to face Allen and Jack gave him another little push. They were getting louder.

"Allen, don't mess with these guys. They aren't worth it," exclaimed Janet.

Dave turned to her. "You shut up."

In less than a heartbeat, Allen stepped on Dave's foot, while giving him a quick punch in the face. He tumbled back. Jack tried to punch Allen, but he easily dodged the punch, grabbed his fist twisting it behind his back, and kicked him in the butt. Jack tumbled across a table knocking it over. They both got up quickly and came at Allen.

Jack picked a beer bottle up as he approached Allen and tried to hit him on the head. Allen grabbed the bottle, swinging it down so that it hit Jack in the groin. He tumbled back in pain and fell over a table. Allen dodged another punch from Dave, grabbed his hand and flipped him on top of Jack. Two onlookers joined the fight for fun and one took a swing at Allen. Allen easily blocked it while punching him in the face. He grabbed the onlooker by the shirt for support and kicked the second onlooker in the knee. Jack and Dave picked up chairs and came at Allen as he flipped the second onlooker on top of the first.

Allen laughed. "If you don't put those down, I'll have to hurt you."

They looked at each other and paused, thinking about it. Suddenly Jack yelled and charged Allen with his chair. Allen grabbed the chair and kicked Jack in the chest. Jack fell like a rock and Allen broke the chair over the first onlooker who had stumbled to his feet and was moving toward him. Allen used one leg of the chair to block a punch from Dave and kicked the first onlooker in the other knee.

A gunshot was heard and everyone but Allen dove for the floor. The bartender was standing in front of the bar holding a starter pistol and was pointing a double barrel shotgun at the fighters.

"THAT'S ENOUGH!! Any more fighting and I'll unload this on someone!"

Allen dropped the chair piece and raised his hands for a second, then lowered them. The others struggled to their feet and started to move away.

The bartender lowered the shotgun. "Dammit. Now... look at this place."

One of the waitresses started to pick up the chairs, and the bartender began to help her. Allen walked over to the bartender, pulled some money out of his pocket.

"Sorry about this." He stuffed the money in the bartender's shirt. When he returned to Janet's table, he laughed at their dazed expressions.

"I'm sorry that got a little out of hand. Are you all right, Janet?"

She suddenly snapped out of her daze. "Oh... I'm fine."

Janet's friend laughed. "I wouldn't say the same for those guys. You whipped their butts."

"That must have been a dream. Where is the real Allen Atkins?"

Allen grinned at her as Jim and Steve patted him on the back.

"Hey, Allen, I hope you don't ever get that mad at me," joked Steve.

"I wasn't mad, except when he told Janet to shut up."

"Well... I know who I want on my side if I ever get into a fight. Damn, I wish I could do that."

"Karate is only used for self defense when you can't talk your way out of a fight."

"Uh... oh, sure. That's what I meant."

Janet grabbed Allen's hand. "Allen, this is my friend Susan Jones."

"Hi."

Admiration was written on Susan's face. "I feel like a damsel in distress who's just been rescued from the evil knight, or something like that. You were great."

"It wasn't that big of a deal. They just had too much to drink."

"Oh, right! Susan, would you mind if I caught a ride home with Allen?"

"Damn, I wish I had said that first."

Janet stood and put her hand on Allen's arm. "How about a ride, handsome?"

"Of course. Susan could you give these guys a ride to the NASA parking lot?"

"I think that's the least I could do."

"Great, let's go," said Janet, tugging on Allen's arm.

As they started for the door, Jim and Steve sat down at Susan's booth. She quickly made some new friends and soon forgot about Janet and Allen.

CHAPTER 7

Phil, Allen, and Ann had finally completed preparations to interview a number of prospective employees while utilizing the new system. A conference room had been converted into a makeshift interview room. One wall of the room consisted of floor to ceiling windows overlooking a park-like setting. Three tables were placed in a «U» configuration with a chair placed in the middle for the person being interviewed. The prototype equipment was set up on one of the side tables, with the laptop situated on the opposite side. Phil was writing on a clipboard, while Allen was typing on the laptop.

Ann was gazing out the window. "Well, it's Monday, and it's a beautiful day. Maybe that's a sign this week will go really smoothly."

Phil scanned the clipboard. "How many interviews do you have scheduled?"

"Almost sixty of the applicants agreed to be interviewed with the system in operation. We should be able to do all of those this week. Then you guys can go back to your jobs, and HR and I can finish the rest of the applicants." She paused to take the top two folders off a nearby stack of employment folders and handed one to Phil. "We're ready if you guys are."

Allen looked up from the laptop screen. "Ann, Phil and I have worked out a sign language to indicate the truthfulness of a response. If an answer is questionable, then you can decide which direction to follow."

"Yes, if the advice value is 75% or greater, we will assume the answer is true and Allen won't do anything. If Allen rubs his nose the probability is 40 to 60% and questionable. Anything lower than 40% is probably a lie and Allen will scratch his head."

Ann seemed to consider that for a moment. "What happens between 60 and 75%?"

"Oh, yeah. Allen will pull on his ear. This is kind of a gray area. The answer is essentially true but there may be something the subject is concerned about. It's sort of like the difference between the 'truth' and the 'whole truth' if you know what I mean."

"I understand."

The legality of using the system still nagged at Allen. "Ann, are you sure this candidate has agreed to be interviewed with the equipment present?"

"Yes! But, I think the first person would agree to anything."

"You know, there isn't much reference data in the database yet, so I would be careful in jumping to conclusions about whether the first few applicants are telling the truth or not."

Phil opened the folder and scanned the first page. "I agree with Allen. I think at this point, you need to base your decision mostly on what you hear and whatever you can prove from other sources, such as previous employers or credit agencies, things you would normally check."

Ann nodded. "I agree as well. This just gives me a different perspective, that's all." She picked up a phone on a nearby table and pressed a few buttons. "Margaret, we're ready now." She hung up the receiver as she continued. "Margaret is my Director of Human Resources."

Margaret Simpson entered carrying a stack of folders. She was dressed in a dark business suit and had a refined, corporate, confident air about her. Margaret had dark hair and eyes and Allen immediately found himself comparing her to Janet. Even Phil tried not to ogle her. She stopped for a second to look at the equipment on the nearby table and then at Ann.

Ann noticed their expressions and chuckled quietly. "Gentlemen, I'd like you to meet Margaret Simpson, our Director of Human Resources."

Allen and Phil shook her hand as they introduced themselves.

"It's a pleasure to finally meet both of you. Ann talks about you all the time."

Allen and Phil glanced at Ann.

"You shouldn't believe everything you hear," commented Allen.

Margaret laughed a little. "Actually, she speaks highly of both of you. Is this the new 'truth detector'?"

"Yes, it is. It's interesting that you would call it that," Phil observed.

"It's an early prototype," explained Allen.

Margaret walked over to the equipment, examining it. Allen returned to his seat in front of the laptop and started typing. Phil sat back down and pretended to look at the file folder in front of him while sneaking a few glances at Margaret.

Ann smiled when she saw Allen glance at Margaret as well. "Are we ready, Allen?"

"We will be in a minute. You can go ahead and send the first person in."

Ann opened the folder in front of her. "Margaret, would you give a brief summary of the first candidate."

"Sure. The position is on the electronic-board assembly line. We generally prefer women in these positions, as they seem to make fewer errors and are able to concentrate on assembly work for a longer time. This is sort of an industry norm."

She picked up a highlighter and marked a few items on the first page. "The first candidate is Shirley Long. A single mother, age 27. She has had some prior work experience on assembly lines but not specifically assembling electronic equipment. We have made some preliminary background checks of her application and found no problems. Is there anything else you would like to know?"

The legality of using the system still bothered Allen. "She knows we have equipment here that will evaluate her stress in relation to her answers and has agreed to it?"

"That's correct."

"Margaret will do the questioning. I don't think there will be any need for us to question her as well. Margaret and I have agreed that on the first few candidates, we will evaluate them without knowledge of how they responded to the questioning. Then we will compare the results."

"That's right. We will also ask a few questions where we already know the answers to help gather the reference data.

"Would you like me to bring her in, Ann?"

"Sure."

When Margaret left, Phil looked questioningly at Ann. "Do you even need to introduce us?"

"Margaret may or may not. I'll leave it up to her. We must be identified if they ask who we are. If we do, we will only give them your names."

Margaret walked in with Shirley Long, the candidate. Shirley was wearing a neat, clean, print dress, obviously her 'best dress' for the interview. Margaret motioned her to the chair in the middle and took a seat in between Phil and Ann.

"Shirley, this is Ann Stevens, Phil Conley, and Allen Atkins."

Shirley smiled nervously as Margaret motioned to each and then continued.

"Shirley, we would like to follow up on a few items on your application. As you know, we are videotaping this interview to help us reach a final decision later. Also we have a stress analyzer here that will help us in our decision on employment." She waited for a moment, but Shirley didn't respond. "I explained this earlier to you and you agreed."

"Yes," she replied, in little more than a whisper.

"We have just a few questions we would like to ask you to get some additional information for your application."

"Okay."

Margaret began with a list of easily verified information, such as her name, address, and social security information.

"How long have you lived at your present address?"

"About two years. Since my husband ran out on me."

Margaret frowned and tried to lead her away from the marital-status type questions that aren't allowed in employment interviews.

"Are you currently employed?"

"I've been out of work for about six months."

Before Margaret could continue, Shirley volunteered additional information not allowed.

"I sold all of my ex's stuff and am still living off that. But that's just about gone, so I'm trying hard to find something. The kids gotta eat, you know."

Allen and Ann were beginning to feel uneasy as the undesirable data continued to tumble out of Shirley's mouth. Margaret was becoming uncomfortable as well.

"Shirley, you have had some assembly line-experience, haven't you?"

"Yes. On my last job, I worked for a brewery on the final-packaging line for a couple of years."

"Well... why did you leave?"

"Because my boss was always trying to hit on me, and I didn't like it."

While the rest were looking even more uncomfortable, Margaret pressed on.

"Why didn't you file a complaint against him with the EEOC?"

"I don't know anything about an EEOC, but I told him once that if he didn't leave me alone, I was going to complain to someone. He told me I didn't have a chance of proving nothing and that I'd be sorry if I tried anything like that. I just figured it was best if I moved on."

Margaret gritted her teeth as she continued. "How did you find out about the opening here? It wasn't advertised?"

"A friend of mine works here. She told me about it."

"What's her name?"

"Wilma Davis. She works in the shipping department."

Ann seemed relieved somewhat. "Wilma's been here almost from the beginning. She's a hard worker."

Shirley piped in before Margaret could continue. "She's been here ever since her husband ran out on her. You can ask her about me. We've been friends a long time."

"Is there anything else you would like us to know that wasn't on the application?" Margaret hoped Shirley wouldn't reply with personal information.

"Just that I'm a hard worker. I wouldn't have left my last job if my boss hadn't been hittin' on me all the time."

Margaret closed her eyes for a moment then looked at Ann and Allen. "Would any of you like to ask anything else?"

They all shook their heads. Phil and Allen were avoiding her eyes.

"Well, then I guess that's about it. We really thank you for answering these questions. I know some of them weren't easy to answer."

"They weren't anything I haven't had to answer before. When do you think you'll know if I got the job?"

"Not too long. I'll show you out."

"No, that's okay. I think I know the way." She stood up. "It was nice meeting you all."

Phil and Allen stood up as Shirley walked out. Ann put her head in her hands. Phil slumped back in his chair as Allen sat down and stared at the keyboard. Margaret wrote in the folder for a moment.

"Well, that went pretty well," she said with a trace of sarcasm. "Did it help you get the kind of data you need?"

Allen sighed. "I guess so. I don't need to tell you everything she said was probably true."

"We guessed that" replied Margaret ruefully.

Allen was thinking of the fifty-nine or so additional interviews to come. "They won't all be like that, will they?"

It had been a grueling week and the strain was showing as Phil, Allen and Ann were all slumped down in their chairs on Friday morning. A clock on the wall indicated 9:00AM as Phil opened the folder in front of him. «God, I›ll be glad when we finish this last group of candidates. I would shoot myself if I had to do this for a living. I think I have a new respect for Human Resources.»

Ann shook her head. "I can't believe we still have 50 more interviews after we're done here. By the way, Margaret won't be here until noon today, so I will be doing the questioning until she returns."

Phil glanced up from the folder. "Let me know if you want me to ask some questions."

The phone rang. Ann touched the speaker button. "Amanda, I thought we agreed on no calls for me during interviews."

"I'm sorry, Mrs. Stevens, but Janet Turner of NASA has called for Allen Atkins several times."

Allen stood up. "Sorry, guys. I have my phone off. I'm not sure what she wants. Amanda, I'll call her back. Phil, would you take my place?"

"Sure."

Allen left through a side entrance, as Phil sat down at the laptop. Ann pressed the speakerphone button again. "Amanda, would you send in Mr. Abrams."

"Mrs. Stevens, Carl Smith has a message from Margaret from you, concerning the next candidate."

"What! All right. Send him in."

As she pressed a button on the phone, Carl Smith, a security guard walked in and handed Ann a note, then stood waiting.

"What's going on, Ann?" asked Phil.

Ann began reading the note aloud.

"Ann. I had a preliminary interview with Gene Abrams and got some bad vibes from him. This may just be my imagination, but I have asked Carl to be present, just in case. --Margaret."

"If he's a potential problem, why bother to interview him any further?" Phil was confused.

"I don't know what Margaret has in mind. Let's see what's going on. Carl, would you bring Mr. Abrams in?"

"Sure, Mrs. Stevens." Carl walked out, then back in with Gene Abrams. Gene's hair was stuffed into a large baseball cap that he forgot to take off. He was wearing a rock concert T-shirt and jeans. His beard was still long, but he had trimmed it some. His mustache had been waxed and twisted into a handlebar.

"Good morning, Mr. Abrams. I'm Ann Stevens, and this is Phil Conley."

Abrams shook their hands. "Pleasure."

"Have a seat. Margaret Simpson should have told you we are videotaping this series of employment interviews to help us make the final decisions, as there are a lot of candidates, and that you agreed to it."

"Yes, she told me."

"She should have also explained to you that we are using a stress-analysis machine to see how you respond to some questions that aren't on the application forms."

"Yes. She told me that too."

Ann quickly scanned the application and some supporting documents. After a moment, Phil saw her frown. Abrams had a multi-page arrest record of misdemeanors and Ann decided to skip the "niceties" and get to the point quickly.

"Mr. Abrams, you have applied for a job in the transportation department. Primarily, this is to pick up and deliver materials locally in the city and surrounding area. It also involves frequent trips to the airport, but no long hauls are required."

"Yes. I'm really a mechanic, but I've had several jobs driving all sorts of trucks."

"As you know, we are a manufacturer of medical equipment. This means we receive a lot of drugs from manufacturers to be tested or used in helping us develop new instruments."

Ann paused, looking for a reaction from Gene, but there was none.

"Okay?" he finally replied.

"The reason I bring this up, is this means you would be responsible for transporting a lot of experimental and prescription drugs. So, we need to know if you have ever been convicted of using any illegal drugs."

Gene didn't appear very happy at having to answer a question like this, and he appeared to struggle in answering it. "I was busted for pot once, when I was in high school. I haven't used anything since then."

Ann noticed Phil scratching his head before he took up the questioning. "Have you ever been involved in the distribution of any illegal drug."

A dark cloud settled over Gene's face, and he reluctantly answered. "No... I don't do that either."

Phil glanced at the screen for second. "How about alcohol? Any alcohol-related offenses?"

Gene suddenly leaned forward in his chair. "Hey! I'm not on trial here, am I? Are you a lawyer?"

"I happen to be a lawyer, but you're not on trial."

Gene's tone turned angrier. "I'm applying for a truck-driving job, not some supervisor job or something."

"We explained why we are asking these questions."

"I know that."

"Then, would you mind answering the question about alcohol-related incidents."

"I've been busted a few times for drinking and fighting in bars. Who hasn't? But that's about it."

Phil glanced at the monitor, then took a more direct tact. "Most of the employees here are women. Have you ever had any run-ins with the police over harassment or domestic violence or some related issue?"

Gene angrily stood up and Carl moved quickly to stand behind him. "I don't have to answer these questions for a job driving a truck. I wouldn't even be around most of the time. What's going on here?"

Ann played the question down. "We are just trying to determine our level of risk if we were to offer you a job. There's nothing personal here."

"It doesn't sound like it!" Gene said angrily. "Then, why is there a security guard here?"

Phil looked up from the laptop screen. "Standard operating procedure. They even have security guards at school board meetings, don't they?"

Gene appeared confused. "Huh?"

"Never mind!"

Ann closed the folder in front of her. "Please sit down, Mr. Abrams. We are almost through."

Gene seemed to cool off some and sat down quickly.

Ann glanced at Phil, but he was staring at the laptop screen. "You can choose to not answer the question if you like."

Gene thought that not answering a question would probably be held against him and that he should try to answer if at all possible. "Uh... I uh... I was arrested for hitting a woman once. She claimed I was trying to steal her money. We got into a fight, and the neighbors called the police. They let me go when she wouldn't file a complaint."

It was Ann and Phil's turn to look surprised. Phil glanced at the screen from time to time. "Why didn't she file a complaint?"

"I told her I'd tell them she hit me first and I was defending myself." Gene almost seemed smug at his answer.

"What started the fight?"

"She accused me of being a thief! That was just too much for the little bitch to say."

Phil and Ann look at each other in disbelief. Ann continued.

"Just two more questions. How did you hear about the job? It wasn't advertised anywhere."

"A friend who knows some people here."

Allen had returned and was standing beside Gene's chair, waiting for a moment to introduce himself as Ann replied.

"I probably know who that would be. Oh, Mr. Abrams, this is Allen Atkins."

Gene stood as Allen offered his hand. "Allen Atkins. Nice to meet you."

Gene nodded as he shook Allen's hand, surprised that Mary's ex-husband was involved. Allen sat down next to Ann.

"Is there anything else that you would like for us to know when we review your application that wasn't on the form?"

Gene looked alternately between Allen and Ann and struggled to answer the question.

"Uh... yeah. I don't do drugs no more. And uh... I appreciate the opportunity to apply for a job. I guess that's about it."

"Allen, Phil, do either of you have any more questions?"

Allen was looking at Gene as if he was trying to remember him. "Have we met somewhere before?"

Gene felt a shiver run down his back. "I don't think so."

"Do you go to kicker bars?"

"Uh... sometimes."

"You sure look familiar."

"I don't think we've met before."

The thought slipped away from Allen. "I guess not."

"That's it, then." interjected Ann. "Thanks for coming in, Mr. Abrams."

Ann, Allen and Phil shook his hand, and he left quickly with Carl trailing behind.

There were almost a dozen people in the outer waiting room. One of them was the HUGE man from the nightclub, Joe Willis, who was waiting to be interviewed. Gene walked out of the interview room with Carl Smith following. Gene and Joe recognized each other, and Gene paused for a second.

Joe snickered and then bellowed "Hey! You still owe me a dance!"

Everyone in the room looked at Joe and then at Gene. Gene scowled and walked quickly out the outer door of the waiting room with Carl close behind. As he left, Joe guffawed loudly.

Phil and Ann were visibly relieved to be rid of Abrams. Allen sat back in his chair. «That was one tough-looking guy.»

"And dangerous," commented Phil.

Ann scanned a copy of a police report attached to Abram's application. "A little too tough for us, I think. I wonder why that arrest for fighting wasn't on his police report?"

"Maybe it just happened, and you have an old report."

"That's possible."

Phil finished typing on the laptop. "By the way, that was about the only thing he said that wasn't a lie."

Ann pressed a button on the speakerphone. "Amanda, would you send Mr. Willis in?"

"Yes, Mrs. Stevens."

The door opened, and a tower of man entered. Joe Willis had tried to make it as a professional wrestler, but he was too honest to take a fall and had reluctantly decided to give it up. Joe was almost seven feet tall, and with a wrestler's weight, exuded a formidable appearance. However, he was well groomed and wearing a nicely tailored suit, which favorably impressed Ann as well as Allen and Phil as they stood up to shake his hand.

"Hi, I'm Ann Stevens, and this is Allen Atkins and that is Phil Conley."

"Hi. Joe Willis," he replied in a deep booming voice.

"Please be seated." Ann scanned Willis' application. "Mr. Willis is here to apply for a job in the transportation department."

Willis leaned forward a little in his chair as if embarrassed to correct Ann.

"Actually, ma'am, I've applied to be a truck driver."

Ann smiled at him. "Yes... Mr. Willis, Margaret Simpson explained to you that we are videotaping the interviews to help us make the final decisions?"

"Yes, she did." Willis smiled. "She sure is a nice lady."

Allen and Phil were thinking the same thing.

"She also explained that we are using a stress analyzer to help us make these decisions?"

"Uh... yeah. What is a stress analyzer again?"

"It tells us how comfortable you are when you are answering the questions. It doesn't change anything, it just measures your reaction to questions."

"Oh, sure." Willis' expression alternated between confusion and agreement.

"Mr. Willis, this report says that you left your last job about two months ago. Can you tell us the reason you left?"

Joe hung his head in embarrassment. "I got into a fight with a guy there."

"Over what?"

They could tell they had hit a sore spot as Joe suddenly became agitated, his voice rising. "He called me stupid. I'm not stupid! I may be a little slow, but I'm not stupid."

Ann motioned him to calm down. "Did you leave on your own or were you let go?"

Willis seemed uncertain of his own answer. "They were giving me a hard time, so I... uh... left."

Phil saw Allen pulling on his ear. "Mr. Willis, are you sure they didn't let you go?"

"They probably would have, but I didn't want to get fired, so I left."

Allen rubbed his nose and Phil pretended to look at Willis' file. "Joe, we contacted them, and they said they had to let you go because you were a real discipline problem."

Ann looked at Phil with a puzzled expression as Willis jumped to his feet.

"That's a damn lie!"

"Joe, calm down. We heard their side, so we just wanted to hear yours."

"I left there on my own!"

Phil motioned him to sit down again. Willis seemed to calm down quickly as he sat down. Ann had read a part of Willis' file. "We also did a little checking on you. You've been arrested a few times for fighting in bars."

"Is that going to keep me from getting a job?" he said meekly.

"Not necessarily. But just out of curiosity, how did you hear about this job? It wasn't advertised anywhere."

"I have a friend in the shipping department. She told me about it."

"That probably would be Wilma Davis," Ann said confidently.

"Yes."

"Well, is there anything else you would like us to know that wasn't on the application form?"

Joe leaned forward in a quiet and honest manner "I'm a real good driver. I've never had a single ticket, and... I'm real reliable. I never miss work unless I'm real sick."

Ann closed the file. "We checked your driving record and couldn't find anything." She looked at Phil. "Anything you want to ask, Phil?"

"Mr. Willis, you have a pretty quick temper, don't you?"

Willis hung his head as he replied. "Yeah. I guess I do."

"It's probably cost you some jobs, hasn't it?"

"Yeah."

"Have you ever considered counseling to help you control your temper?"

"They tried to get me to do that at my last job."

"If you had, you might still be there."

Willis seemed to consider that before he answered. "You're probably right."

Ann opened Willis' file and scribbled in it. "Joe, if you were offered a job and it was required that you get some counseling for your temper as a condition of employment, would you agree to it?"

He nodded slowly. "My life's not too good right now and my temper is mostly to blame. Yeah, I'd do that."

"Good. That's all the questions we have. You'll be hearing from Margaret pretty soon. It was nice meeting you."

Joe lumbered to his feet. "Thanks for hearing me out."

Ann and Allen stood up and shook Joe's hand. When he had closed the door, Allen stood up and stretched. Ann turned to Phil. "What's going on? We didn't contact his last employer."

"Allen indicated he wasn't telling the truth, so I just followed up on the question, just like I would hope we would do in the courtroom."

Allen rubbed his eyes, picked up an empty coffee cup, and threw it into the trashcan. "Everything he said was probably true except the part about being fired at his last job. And that was borderline. He probably isn't sure if he did leave before they terminated him."

Ann scribbled something in Willis' file. "He seems like a big teddy bear."

"He does have a temper," commented Allen.

Phil closed Willis' file. "He does, but I think, with a little counseling, he would make a good employee."

Ann took the top two folders off a nearby stack of folders and handed one to Phil. "Okay. Let's do one more then take a coffee break."

"By the way, Allen, is there a problem at work?"

"No. Janet called to let me know that a project that I was working on in the planning phase has been approved."

"That's great! Congratulations!" Ann said as she shook his hand.

"Yes, but she also said my technician told her a neighbor of mine called to say that Mary's in the hospital and asked that I go see her."

Phil and Ann exchanged concerned looks. Phil put his hand on Allen's shoulder. "Mary's in the hospital! I'll go with you if you don't mind. I haven't seen her in a long time. Did Janet say why she's there?"

"The neighbor didn't tell her."

"When are you going?"

"I was thinking about going tonight."

"That's fine. We can go after we finish here."

Ann was apologetic. "I would like to go, but I have a business meeting with a potential investor that I can't miss." She looked at her watch. "Let's take a quick coffee break then regroup."

Ann, Allen, Phil and Margaret spent the afternoon reviewing the week's interviews.

"It was a tough week," commented Allen, "but we now have a ton of data in the database, and I think we probably are much nearer to the three-sigma correlation."

Phil wearily rubbed his eyes. "Does that mean we're ready for LA in a few weeks?"

Allen stared at a printout of the correlation data. "I would say yes."

"That's great! I'll go tell Adam we're coming for sure." Phil took out his phone looking for Adam's telephone number.

Margaret shuffled a stack of folders into a neat pile. "This sure has been an interesting week, but we still have over 50 people to interview next week, Ann."

"I think you can handle the routine stuff without me. I'll come in when you think it's necessary." She glanced at the equipment that had been packed in boxes. "I can't wait to see this system operating in a courtroom.

Margaret shuffled the folders from the day's interviews into a stack. "I wish I had one of them all the time."

Ann laughed. "You'd be dangerous then."

Margaret stood up and picked up her folders to leave. "It was a pleasure to meet you."

Allen shook her hand. "Nice to meet you. I think you did an excellent job this week."

"I agree." Phil shook her hand also.

"I would like to hear how the system works on a real case. Please let me know."

"We will, for sure."

Allen sighed loudly when Margaret closed the door behind her. Ann chuckled softly. "In case you're interested, I don't think she's going with anyone right now."

Allen looked up from a notepad he was writing on and smiled.

"Well, guys," Phil said, "the easy part is over. Now we get to see how it'll work in real life."

CHAPTER 8

Phil and Allen were looking for Mary's hospital room number when they ran into Joyce Wilson, a former neighbor of Mary and Allen's. She was happy to see both of them.

"Allen! I'm so glad you could come." She hugged Allen and then Phil. "Phil, I haven't seen you for a long time. How have you been?"

"I'm fine. We have been really busy the last few years. You know how it is, more cases with no increase in staff."

"It's the same thing where I work too." She took Allen's hand. "Allen, I'm really glad you and Phil could come. Mary is really depressed right now."

"How is she? I didn't hear what happened, just that she was here."

Joyce suddenly seemed embarrassed. "Her boy friend beat her up."

Phil and Allen stared at each other in shock.

"I'm really sorry to hear that. I didn't even know she had a boyfriend," replied Allen, "Is she okay?"

"Physically she is. The police came before things got too bad." Joyce paused to look into Allen's eyes. "Emotionally, she's a wreck."

"That's understandable. Did the police arrest the boyfriend?"

"They took him to the police station, but they had to let him go because Mary wouldn't file a complaint against him."

Phil shook his head, obviously angry. "Why not? I just don't understand things like that."

"I think she's afraid of him, but I'm not sure."

"Can we see her now?"

"Sure! Just remember she is really down right now."

"We'll be careful."

66

Joyce entered, followed closely by Allen and Phil. "Mary? Guess who's here."

Allen was shocked at the bruises and cuts on Mary face and her black eye.

Mary's face lit up at seeing him. "Allen! I'm so glad to see you."

"I didn't know you were here or I would have come sooner." He sat on the bed, leaned over and hugged her. "How are you?"

"Much better. I think I can go home tomorrow."

"That's great!"

Phil stood next to Allen.

"Phil! It's great to see you. I didn't expect to see you here."

Phil held her hand. "I didn't expect I would see you here either. How are you?"

"Recovering, I guess."

"What happened, Mary?" Allen asked softly.

Mary looked away from him for a moment before answering. "I started going out with a guy I met at a night club. I don't even know why now." She wiped a tear away as she continued. "Last Wednesday I received the final settlement check from the sale of the house. He wanted me to loan him most of it for something he said he needed." She paused again, staring at Allen. "I can't even remember what it was now. We got into an argument over the money, and he just went berserk. He started breaking things in the apartment and then started hitting me. I guess the neighbors called the police, because they came right away."

Phil turned away, too angry to look at her as she continued.

"I guess I was lucky they came, or he might have really hurt me." She started to choke back tears.

Allen felt a rush of anger that was quickly replaced by sorrow as he looked away. "I'm so sorry."

"I guess I had this coming, for what I did to you."

"Don't be ridiculous. Nobody deserves this."

Phil fought off a wave of anger. "Mary, why didn't you file a complaint against him?"

"I was afraid he would do something even worse when he got out. He really is an okay guy, until he starts drinking. Then he gets mean."

"I hope you aren't going to see him anymore." Allen felt his anger draining away.

Mary shook her head as she put her hand on Allen's. "I'm so sorry about the divorce and all the lies, Allen. That was all Jason's idea. He said that almost any judge would believe that, and there wasn't any way to prove it wasn't true."

Allen glanced at Phil. "Maybe there will be some day. Mary, is there anything I can do for you?"

"I would like to help too," said Phil.

"No. Not really." She was silent for a moment as she summoned strength to ask him. "Allen, I was so wrong about you. Is there any chance you could forgive me, and we could start over somehow?"

The answer was difficult for Allen. "You know I still have strong feelings for you. I will always love you in a way. I forgave you a long time ago. I guessed that you must have been desperate to want to get rid of me that much."

She tried to smile at him. "But..."

"But I have started going out with someone too, and I have to give it a chance to work. It wouldn't be fair to her if I didn't."

Phil and Joyce had moved slowly to the door to leave them alone.

Mary's disappointment showed, and it hurt Allen when she replied. "I understand, Allen. You always were a fair person. I guess I just didn't appreciate you." She managed a weak smile.

"Thanks for that. If things don't work out, I'll be knocking on your door."

Mary looked away and then back at him. "I think I may go back and stay with my parents for a while. It may help me get my life back together."

"That's a great idea. When you're better, will you give me a call? I would like to hear from you."

"Of course. Allen?"

"Yes?"

"Will you do something for me... before you go?"

"Sure!"

"Give me one more hug."

He leaned over and hugged her for a moment. "I'd better go. Promise me you'll take care of yourself."

Mary's reply was barely audible. "I will."

Allen picked up her hand, kissed it, and then stood up. "Do you want me to get Joyce for you?"

"Sure. Thanks, Allen."

Allen opened the door and stopped to look back at Mary. "I'll see you."

He walked out and a moment later Joyce walked in. Mary was crying quietly as Joyce hugged her.

CHAPTER 9

Allen and Janet were sitting in a corner booth of their favorite Mexican restaurant, having spent the day together shopping, picnicking on the shore of Clear Lake, playing tennis, and finally having a quiet dinner together. They were holding hands, and Janet had her shoes off, playing «footsie» with Allen when the waiter arrived with gourmet coffee.

"Would you like to see the dessert menu?"

Janet was paying too much attention to Allen to look at him as she replied, "No, thanks. I couldn't eat anything else."

"I'll pass too. Could we have the check?"

"Yes, sir. I'll be back in a minute."

"So... what time does your plane leave tomorrow?"

"Early... around 7:30 in the morning."

"How long do you think you'll be gone?"

"A month or so. Two months, tops."

"Two months is a long time. You'll probably meet some blond nymph out there and forget all about me."

Allen laughed. "I think the chances of that are slim and none."

"Ha!"

Allen started to reply, but the waiter returned with the bill.

"Thank you very much. Come back and see us real soon."

As he left, Janet put her other hand on Allen's. "I suppose you have to hurry home now and finish packing."

"Actually, I finished last night."

"So, then... how about coming over to my place for a while."

"It's pretty late. Tomorrow's a regular workday for you. I wouldn't want to keep you up."

Janet gazed into his eyes. "Yes, you would."

Allen grinned at her. "Then, what are we doing here?"

They stood up quickly. Allen left money for the bill, and they almost ran out to his car.

As Phil pulled into the parking lot of Joanne's apartment building, he was wondering how she would react to his one-or-two-month-long absence to help his old friend, Adam Daniels. He sensed she was not very happy when he told her on the phone. But she had invited him to dinner, and he was hopeful he could make amends when he returned. He picked up the flowers he had bought for her and walked briskly up the stairs to her apartment. He took a deep breath and rang her doorbell, wondering how she would react.

"Who is it?" she asked through the door.

"It's Phil." He heard her make a comment he couldn't quite make out.

When Joanne opened the door, she was wearing only a short negligée. It took Phil's breath away. He was in a daze and a few seconds passed before he handed her the dozen roses. She smelled them quickly and tossed them onto a small table near the entry door. She took his hand, pulled him inside and kicked the door closed.

"You may be leaving tomorrow but you're mine tonight."

She had her arms around him and was kissing him passionately for a moment before he came to his senses. He picked her up and carried her into the bedroom.

It was a little past 3:00AM as Allen pulled into his apartment complex. He kept thinking about Janet and the intimate moments they had enjoyed. A few months ago, he wouldn›t have believed it would ever happen, yet now that it had, it seemed that something was missing. He hated to say it was «passion,» but he couldn›t think of a better description. Perhaps he still wasn›t over Mary and was somehow comparing Janet to her, or maybe Janet wasn›t over her first husband. Whatever it was, it left him wishing there had been something more than the physical enjoyment of the moment. He sighed. Maybe it would be better next time.

The sun was just peeking over the horizon as Ann, Allen and Phil settled into their large comfortable seats and ordered drinks. Ann was sitting next to the window and, ever the businesswoman, she pulled the window shade down and took a stack of papers out of her briefcase and started skimming through them.

Allen pushed his seat back and lifted the footrest. "First class! And not even a frequent-flier upgrade."

Phil laughed and put his briefcase under the seat in front of him. "I told you money is no object for this guy."

"How come he has so much money? What does he do?"

"He's an investment guru. He used to be an account manager for one of the big money funds before he went out on his own. I think he invested everything he could get his hands on and made out really well. He must be worth four or five hundred million now."

Allen stared at him in amazement. "HUNDRED million! I don't think I'll ever see any serious money."

Ann put her papers down. "Guys, I hate to change the subject, but I assume you were able to test the third prototype."

Allen opened his briefcase as he replied. "Yes, it seems to produce exactly the same results as the second. By the way, that miniaturization you did was first class. I didn't think you would be able to put everything in a salesman's briefcase like that."

"It comes from a lot of years of building prototypes." She pulled a small calculator out of her briefcase. "By the way, what was the final cost of your equipment?"

Allen pulled two papers out of his briefcase and gave one to Phil and the other to Ann. "I wasn't too far off. Also, we lucked out on the spectrophotometer. The company that made the miniaturized version for NASA had made a few extra hoping to sell them a few more in the future at an elevated price. I was able to get one for about the same price as a standard model. By the way, I don't think you could have put a standard unit in the suitcase."

"How about the artificial Intelligence software?"

"I've entered all the interview data."

"That's great! How about the latest version of 'Sherlock'?"

"Yes, I have version 3.0 with me. You'll see quite a difference. I think he has some of Tom and his wife's personality now."

"I need to bring you up to speed on the case." Phil handed Allen a thick folder. "Let's discuss it when you've read what you need."

Allen stared at the huge folder for a second. "How long IS this flight?"

As the jumbo jet made its final descent into Los Angeles International, Phil leaned over to look out the window. Ann kissed him and he jerked back. «Don›t, Ann!» he whispered.

"Why not? We're still friends, aren't we?"

"Yes, but...."

"We were a lot more than that once. I almost hooked you; you know."

"That was a long time ago."

"Sometimes it seems like yesterday."

"You've been married to Dan for eight years!"

Ann started to run her hand on his face. "We tried to make it work. Dan's an okay guy, but his career is more important than his family. And... he's not you."

Phil pulled her hand away. He glanced at Allen who was nodding off. "We need to work together on this case, Ann. Don't make it difficult for me. Please!"

She sat back in her seat and smiled at him. "I'll be a good girl."

CHAPTER 10

Phil, Ann and Allen walked out of the jet way and into the gate area where Adam Daniels and a companion were waiting. Adam was a little younger than Allen would have imagined, to be so wealthy and famous. Allen guessed Adam was in his late 40s, slender and with a classic gray streak in his temples, was very distinguished. Adam's companion, with closely cropped hair and an extremely muscular build, presented an image of a professional football player wearing a suit. Phil was a few steps ahead of the others and walked over quickly to shake hands. Allen smiled as they patted each other on the back. Phil quickly introduced them to his old friend.

"Adam, this is Ann Stevens."

Adam seemed genuinely pleased to meet Ann. "Of Stevens Specialty Equipment? It's a pleasure to meet you too. It may come as a surprise to you, but I used to own some stock in your company."

It was Ann's turn to look surprised. "Really? I thought we were a well-kept secret."

"Sometime while you're here, I would like to talk to you about your company."

Ann's surprise showed. "Oh! Sure!"

"Adam, I'd like you to meet Dr. Allen Atkins, the NASA physicist and chemical engineer I've been telling you all about."

Allen and Ann observed that Adam seemed surprised and stared at Allen for a brief moment before he shook Allen's hand. When he did, he seemed apologetic.

"Oh... a real pleasure to meet you, Dr. Atkins. Phil usually doesn't speak so highly of scientists."

"I'll bet. It's a pleasure to meet you too. Is there something wrong, Mr. Daniels?"

"I'm sorry, it's just that you look so much like someone I know." Adam then introduced his companion. "Everyone, this is my assistant, Roy Williams. If each of you will give him your baggage tickets, he will collect your luggage for you. I understand you have brought a lot of equipment with you."

"Almost everything we need," replied Allen. "There are a few things we will need to purchase out here."

"It would be my pleasure if you all would stay with me while you are out here."

"We wouldn't want to impose on you like that," replied Phil.

"It really wouldn't be any problem. I have plenty of room. And it might make it easier to review the results of your testing or to discuss strategy with Ross' lawyer."

"We had planned to stay at the Hilton downtown near the courthouse. It's a two-minute walk and we may need to visit there often."

"That's fine. But if you change your mind, just let me know."

Allen, Phil and Ann fumbled with the baggage receipts, handing them to Roy.

"Roy, please take their luggage to the Hilton, check them in, and then bring all of their equipment to the house."

"Right, boss."

"Would you all like a drink now, or would you rather come to my house to rest and talk for a while."

"I'd just as soon relax for a while."

Ann turned to Adam. "That's fine with me."

"Sure, let's go."

They walked to the exit marked "ground transportation" where Adam's chauffeur and limousine were waiting.

When Brian Limpanatti reached the baggage claims area, he took out a cellular phone. He was dressed as a tourist as he tried to blend into the background. He carefully positioned himself behind a building support column to watch Roy Williams and a skycap gathering luggage onto a cart.

"Boss! This is Brian. No, I couldn't get close enough to hear all of their names, but one of them is Ann Stevens or something like that. One guy introduced the others, so I didn't hear his name. He must be the friend of Daniels. The third guy was introduced as doctor somebody."

He paused to take a quick peek at Williams and the skycap.

"I'm sorry, but there were just too many people to get closer. And I think I recognized a former cop who's a private dick now. No, he didn't see me. What do you want me to do now? I heard them say something about staying at the Hilton. Do you want me to follow Daniels' flunky? He's collecting all their baggage now, so he might be going to their hotel. Okay, Okay. I gotta hurry now, he's getting ready to leave... Bye."

Brian casually followed Roy Williams and the skycap as they pushed a luggage cart out of the baggage area.

In the limousine, the talk quickly turned to Ross› trial. Allen was still holding the folder on the case. «I hope we can help your son. Phil and I have been reviewing all the information you provided.»

"I really appreciate you taking time off from your work schedules on such short notice. This has been a real nightmare for my wife and me, as well as for our son."

Allen opened the folder and thumbed through it. "Is it true the person murdered was the son of a mob boss?"

"Yes. I think he was kind of wild and somewhat of an embarrassment to his father. But his father overlooked all that, and now he's on a vendetta to put someone away for the loss of his son."

Phil rubbed his chin. "Are there any special security concerns we need to be made aware of?"

"Your safety won't be issue. You probably didn't see them, but I had several security people at the airport to make sure there wouldn't be any trouble."

"That must mean they know about us!" Allen replied anxiously.

"They have a lot of informers everywhere. I'm certain they know you're coming out here to help with the case." Adam seemed unconcerned. "They probably don't know anything about the nature of your equipment or the testing you're going to do."

Phil's concerned showed. "How's Ross holding up?"

"As well as can be expected. We've talked about your offer to help and a little about what you're going to do, and he's anxious to meet you all."

Ann was sitting next to a miniature bar and made herself a stiff drink. "When can we meet him?"

"I think tomorrow. I'm still working with the police to get you cleared to talk to him. The detective in charge of the case just wants to talk to you first."

"When can we meet Ross' lawyer, Ryan Hughes?" Phil was worried about the defense's apparent lack of progress. "We have a lot to talk about before we do anything related to the case."

"He's currently waiting at my house. He's very interested in the testing you're going to do. Ann, Phil told me you have come up with some improvements to the system."

"Not really. I just found a way to miniaturize some things. I have some interest in a part of this system for some possible medical applications."

"Once I know some more about all this, I may want to talk to you and Allen about investing in the system, either the lie-detector aspect or the possible medical application. I've been lucky a few times to get in on the ground floor of new technologies. Allen, would you mind giving me a brief overview of how your system works."

"Sure," he replied.

Allen spent most of the trip to Adam Daniels› mansion explaining the operation of the lie-detector system. «So, you see, the final output of the system is merely the 'advice' number that is a calculated value based on a composite of the different methods."

Adam was rubbing his chin thoughtfully as he stared out the window. He glanced back at Allen. "And the interview testing you did at Ann's, that helped add to the database of normal responses?"

"Yes, I think it helped a great deal." Allen closed the large file folder he had almost finished scanning. "I hate to change the subject, Mr. Daniels, but I reviewed some of the testimonies and there seem to be some inconsistencies in your son's statements."

Adam stared at him. "What do you mean, 'inconsistencies'?"

"I'm not a criminologist or a lawyer, but I would like to ask your son a few questions."

"Do you mean you think he's hasn't told us the whole truth?" Adam's reply was laced with concern.

"I don't know until we test him. Would that be possible? Could we set up the system at the police station and ask Ross some questions?"

"I'm pretty sure it's okay, but I'll have Ryan find out. Oh... we're here."

Ann, Phil and Allen were amazed at the size and grandeur of Adam's mansion. The architecture of the mansion, located on the top of a small hill, was similar in style to the classic southern plantation, with white columns supporting a veranda that encircled the house.

Adam saw their expressions. "I didn't build it. It was built by a movie star, and I just fixed it up a little."

They stared at him as the limousine pulled through a security gate, climbed a circular driveway and glided to a stop in front of the huge palatial mansion in Bel Air.

As they exited the limousine, Phil noticed several men standing at strategic positions around the house and gardens. Adam turned to him.

"Please take the others inside, I'll be there in a minute."

Two security men approached Adam as he stood beside the limousine. As they started toward the front door, Allen commented to Phil, "This must be the security he was talking about."

"I hate to admit it, but this makes me somewhat uncomfortable," he replied. As they neared the front door, Ryan Hughes emerged to meet them. Allen was surprised at the apparent youth of Ross' defense attorney. He would have guessed Ryan to be in his mid 20's, with almost a boyish appearance. Ryan walked toward them and briefly paused with a surprised expression, before he introduced himself.

"Uh... hello! I'm Ryan Hughes. It's a pleasure to meet you all." He shook Allen and Phil's hand.

"I'm Phil Conley. Pleasure to meet you too."

"Ann Stevens," Ann said, shaking Ryan's hand.

"Allen Atkins. Pleasure."

"Dr. Atkins?

"Just Allen."

"Sorry. Allen?"

"Yes?"

"I don't know if Mr. Daniels told you, but you look incredibly like his daughter's ex-husband."

"He didn't say that. He did say I look like someone he knows."

Allen and Phil's attention was immediately drawn to Adam's daughter, Alicia, as she walked over to her father who was talking to what appeared to be the head of security. Alicia was an extremely attractive blond in her mid-twenties. She appeared to have come from a tennis match. Ryan continued his discussion with Allen.

"The resemblance is remarkable. But we shouldn't be standing outside. Let's go in and relax and have a drink."

Phil whispered to Ann. "I wonder why we shouldn't stand outside?"

As Ryan started to lead them into the house, Adam called to them. "Wait, I would like you all to meet my daughter."

Alicia was in an animated conversation with her father as they approached.

"...this is such a waste of time. I really have a lot to do."

"Just humor me."

When Alicia saw Allen she stopped in surprise. "Oh!"

"Gentlemen, this is my daughter, Alicia."

Phil was the closest and he stepped forward to greet her. She smiled at him, then hugged him. "Hi, Phil. Great to see you again."

Phil turned to his companions. "I'd like you to meet some old friends of mine. This is Ann Stevens."

Alicia stepped forward and shook her hand. "Pleasure to meet you, Ann."

"And this is Dr. Allen Atkins."

Alicia had a look of disbelief on her face. She weakly held out her hand to shake Allen's. "Hello, Dr. Atkins."

He had to reach over to shake her hand. "Please call me Allen."

"Oh... sure," she replied absently.

Her father tugged on her sleeve to get her attention. "Well, was I right? Doesn't he resemble Tom?"

"Yes... you certainly could say that."

"I've heard that there are many people in the world that resemble each of us," replied Allen.

"Perhaps. But you could be his twin. Alicia, you can get back to whatever was so important now."

"Oh... that's okay, dad."

Adam began to walk to the house, and they followed him. Alicia quickly moved to walk next to Allen. "I think I would like to stay and learn about this new invention of yours, Allen. Do you think it can help Ross?"

"I don't know. We've done some preliminary testing with it, but we haven't actually used it in court yet."

Ryan fell into step with them. "Well, I for one would like to learn all about this new lie detector, and especially your take on this from a legal aspect, Phil."

Adam made an announcement before Phil could reply. "Before we get too deeply into this, I have arranged for a caterer to meet us in the dining room. Why don't we continue this conversation over lunch."

Everyone indicated agreement as they followed Adam inside.

CHAPTER 11

Brian Limpanatti waited impatiently with Mario Tartela outside a hotel room. There was a housekeeping cart just outside the door and some passionate lovemaking sounds could be heard through the slightly open door. Mario and Jack occasionally laughed softly and whispered to each other.

Mario briefly peered into the room, then looked at Brian. "Why don't we just knock her out and take the damn key?"

"You know we can't take the risk of getting the cops involved. Plenty of people have seen us around here. How much longer?"

"Who knows? Some things can't be rushed."

They both laughed quietly. There was a brief period of silence and after a moment, Juan Ortega opened the door, still zipping up his pants. He glanced back at Maria Lopez who was sitting up in bed, holding a sheet up in front of her. He stopped and looked back at her. "Oye! Hasta Luego!"

He blew her a kiss, and she giggled. As he closed the door behind him, he reached into a pocket and pulled out a card key, tossing it to Mario.

Mario stared at it. "Are you sure it's the right one?"

"Of course, it is. I seek perfection in all my work."

The three men laughed softly.

Mario looked at his watch. "Come on, let's go. It's almost 7:00." They hurried to the elevator.

As Phil, Ann, and Allen were leaving for the evening, Alicia had her hand on Allen's arm as they walked out to the waiting limousine. Her sudden attention to Allen was not missed by her father, or the fact she had changed into an evening dress and was wearing lipstick. Alicia rarely wore makeup.

"Allen, I think your idea is great. I hope your system can help Ross. I also hope you can come by tomorrow evening and let me know what you find out from Ross."

"If it's not too late, I'm sure I can."

"It doesn't matter if it's late. You know dad and I really want to know what really happened."

When they arrived at the limousine, Alicia gave Allen a quick hug, whispered good-bye, and stood waiting next to her dad as the others entered. Adam and Alicia remained for a moment, watching the limousine leave through the gate, which closed automatically after them.

"You were attracted to Allen."

"Was it obvious?"

"Only to a father."

She kissed him on the cheek.

"Does it have anything to do with the fact that he looks so much like Tom? I know you still have some feelings for him."

"It may have been that at first, but after being with him for a while, there is something about him that... I don't know. Maybe he just seems to be hurting inside, and I need to comfort him or something. I just don't know."

"You know he's divorced?"

"Yes. He told me."

"Did you know it was only about six months ago?"

"No, I didn't know that. Maybe that's what it is. He's still not over her."

"Just be careful, Alicia. Don't try and fill a void in his life."

"I know what you mean. I don't want that either. Do you think this new lie-detector method can help Ross?"

"I pray it can. We need a miracle right now."

"Maybe this is it." Alicia shivered in the night breeze. "Brrr. Let's go in."

Adam put his arm around her, and they walked into the house.

As the limousine left the Daniels estate, Ann saw a worried expression on Phil's face as he stared out the window. «What's wrong, Phil?»

"Doesn't the makeup of the defense team bother you a little?"

"What do you mean?"

"Well… the lead attorney has been practicing less than five years, the other attorney is right out of law school, and then there's one paralegal and a few detectives."

"When you describe it that way, yes!"

Phil rubbed his eyes wearily. "Allen, wouldn't you even agree that Ryan might be in over his head on this case?"

Allen wasn't sure what he meant. "Well, he does seem a bit young to be handling such a high-profile murder case. Is that what you mean?"

"Not exactly. I reviewed the case privately with Adam shortly after we arrived and expressed some concerns about the defense team's lack of effort at proving Ross' innocence."

Ann frowned. "I would have to agree. Adam can certainly afford whatever help Ryan needs, but there doesn't seem to have been much done. The defense's case at this point seems pretty weak."

"It's almost non-existent."

"What did Adam say?"

"He said they had hired some additional detectives, but they hadn't come up with very much yet. He did seem worried at their lack of progress."

"Why is Ryan the lead attorney on the case?" Ann wondered how a young, apparently inexperienced, lawyer could have been chosen as the lead attorney.

"I asked Adam that. His personal lawyer suggested Ryan's firm. They first proposed using their top-notch defense attorney, but he was almost killed in a freak automobile accident and Ryan somehow managed to wrangle the job from the next most likely candidate. Adam merely accepted the law firm's recommendation. He did concede that Ryan didn't have much experience for such a high-profile case. I guess at first, he just trusted that the firm knew what they were doing. I think this is why he was so grateful for our offer of help."

"If Ryan wasn't performing, why didn't Adam get another lawyer for Ross?"

"Ryan kept assuring him they were doing everything possible to find the truth. Adam also relies on his daughter for advice, and she seemed to think Ryan was trying hard to help her brother."

"I guess Adam thinks Alicia is a good judge of character." Allen saw Phil smile and shake his head. "What's wrong?"

"I just don't know what it is that you have, that seems to attract beautiful women to you like flies."

Ann and Allen laughed.

"She's no dumb blond. Did I mention she was on the cover of several magazines?"

Allen was suddenly interested. "No. Which ones?"

"Money, Today's Investor, and a few other investment magazines."

"What?"

"When she finished her MBA at Wharton, she started to work for her father. He was so impressed, he turned one of his mutual funds over to her, and in a few years, her fund was beating all its competition. The magazines showed up for interviews and were so impressed they put her on the cover."

Ann chuckled at Allen's surprised expression. "So... pretty and smart, Allen."

"She probably isn't over her ex-husband yet, and I just happen to look like him. That's all there is to it."

"And cows can fly," Phil laughed loudly.

"It doesn't matter. I have someone pretty special back home."

"Janet IS a babe, but she doesn't have a couple hundred million in the bank."

"That's pretty crude. There's a lot more to life than money."

"You're right. You know, Allen, if anyone else but you had said that I'd say they are full of it. But you've turned down some real money-making opportunities to do what you like to do at NASA."

"I love my job."

Ann nodded approvingly. "Not too many people can say that and really mean it."

They rode on in silence.

Brian Limpanatti peeked from behind a huge fan palm in the hotel lobby as Phil, Ann and Allen walked up to the hotel desk to obtain their keys. He backed up quickly into the shadows, pulled out a walkie-talkie and began speaking softly.

"Mario?"

"What?"

"They're here. Get going."

"I need a few more minutes."

"You don't have it. They'll be up there in two minutes."

"Can you stall them?"

"Are you out of your mind? I can't let them see me!"

"All right."

Brian put his walkie-talkie away as he exited though a door marked "Housekeeping."

Mario was in the bedroom portion of the suite closing a suitcase when there was the sound of a card key in the front door. He could hear voices as the door opened. Mario quietly closed the bedroom door leaving it open a crack to listen. He pulled a gun from his waistband and crouched down as Phil and Allen entered the front room.

"So... what's your professional opinion at this point? Do we have a chance of helping him beat these charges?"

"I'm optimistic. I hope we can resolve some of the apparent conflicts in his story tomorrow."

"Excuse me for a minute." Allen walked quickly to the bathroom that was in between the front room and the bedroom. Phil headed for the balcony. He opened the curtains and the balcony door and stepped outside. Mario quickly opened the bedroom door and walked quickly to the front door. He was closing the door behind him as Phil turned around and started to walk back inside. He was just in time to see the front door closing.

"Allen! Where are you going?"

Allen came out of the bathroom door. "What do you mean? I'm not going anywhere?"

Phil whirled around. "What the... Allen, did you just open the front door?"

"No, I was in the bathroom."

Phil walked to the door, opened it and looked up and down the corridor, but he couldn't see anyone. He closed the door. "I don't get it. I'm sure I saw the door closing."

"Maybe the maid came in to turn down the bed and realized you were here."

"Maybe. But they usually knock first, don't they?"

"I don't know. Forget it."

Phil yawned and started to walk to the door. "It's been a long day; I think I'm going to bed. I'll see you in the morning for breakfast."

"Around 7:30 in the lobby?"

"All right."

Phil walked out the door and Allen followed him to peer up and down the hallway one more time. He closed the door and put both deadbolts on. He walked to the bedroom and picked up a suitcase and put it on the bed, opened it, and started to take out some pajamas. He stopped, picked up a travel bag and noticed that it was unzipped. He stared at the deadbolts on the front door from his position in front of the bed, then picked up a phone near the bed and dialed a number.

"Phil? Listen, would you do me a favor? Before you go to sleep, look at your things in your suitcase and see if anything is unusual, missing, or just in the wrong place. We can talk about it at breakfast. No. You don't have to do it now, just let me know in the morning, okay? Thanks. Goodnight."

He hung up the phone, and picked up the travel bag again, starting to look through it.

Mario quickly joined Brian and Juan in a supplies closet. Brian peeked out the door to make sure no one had seen them enter the closet. «Did you give the card key back in her?»

"Of course! I said I found it outside the door, near her cart."

"I checked out this Ann Stevens. She had a lot of papers in her briefcase, but they didn't have anything to do with the trial. I did get her address off the luggage tags. I called Maury in Houston, and he's checkin' her out."

Juan pulled a notepad from his pocket. "I checked out Phil Conley. He must be the lawyer. He had a big thick folder of trial stuff in his briefcase."

Mario pulled out his notepad. "My guy works for NASA, but I didn't find anything about what he does. He did have a lot of small tools and some weird electronic shit in his luggage. But, you know, he didn't have a lot of clothes. So, maybe they aren't stayin' very long."

Brian suddenly had an inspired look. "You know, Daniels' guy picked up a bunch of aluminum suitcases with the luggage. Did you see any of that?"

Juan shook his head. "Nothing like that."

Brian had a bad habit of blurting out whatever was on his mind without thinking very much. "I bet they took that stuff to Daniels' house. We'll have to look through it there somehow."

Mario's mouth dropped. "Are you nuts? Daniels has guards all over the place. We couldn't get close to it, even if we wanted to."

Juan scratched his head. "Why don't we call the boss? We could ask him what he wants us to do."

Mario and Jack looked at each other as if to say, "Why didn't I think of that?"

CHAPTER 12

The reception area was crowded with people waiting for visiting hours to see relatives and friends. Several deputies were in the room and two were behind an information desk along with two county employees in civilian clothes with nametags, as Ann, Phil, and Allen entered carrying two aluminum cases. This immediately drew the attention of several deputies, who instinctively unbuttoned the strap on the holsters and put their hands on the handles of their guns. Adam and Ryan were standing next to Detective Dave Johnson providing him with additional information on the testing they wanted to do.

Dave Johnson was in his late forties, slim, and ruggedly handsome. His tanned face testified to the time he spent in outdoor activities. He immediately noticed the reaction of the deputies to the suitcases they were carrying, and quickly spoke to them above the din of the crowd. "It's okay, gentlemen. They're here to see me."

The deputies relaxed and buttoned the strap on their guns, returning to their business. Adam motioned them over. "Over here, Allen."

They threaded their way through the crowd to greet Johnson, putting down the suitcases.

"Ann, Allen, and Phil, this is Homicide Detective Dave Johnson. He's handling the case. Dave, this is Ann Stevens, Allen Atkins, and Phil Conley."

They all shook his hand. "Pleasure to meet you. Let's go back to my office. Your equipment will have to be searched, of course. Standard operating procedure."

Allen nodded in agreement. "Of course."

Ann, Ryan and Detective Johnson waited by a door to the office area for the others as their suitcases were inspected.

Detective Dave Johnson shared an office with two other detectives. The office was crowded with three desks, several file cabinets and a dozen chairs. Personal computers on each desk competed for space with piles of paper. The other detectives were not in the office as they entered, and Johnson motioned them to the chairs. He gestured to the corner of the office. "Please help yourselves to coffee."

Ryan poured himself a cup of coffee and sat down near Johnson. "Thank you. We appreciate your agreeing to let us videotape our interview with Ross."

"It's a bit unusual but I don't see any harm in it. I understand you're also trying a new software package that helps solve crimes."

"Yes, it's based on an Artificial Intelligence software package. We hope it will help point us in a new direction, and perhaps help you catch the real killer."

Johnson smiled and sat back in his chair. "Nothing would make me happier. You know, we have several software packages that we use to help us in our investigation. I would be interested in comparing your system to ours."

"I'm sure we can arrange a demonstration sometime."

"It's an early prototype, though," said Allen.

"You said we could use one of the interrogation rooms?"

"Yes. You also understand that there must be an officer present at all times."

"Certainly."

"I'm sorry I won't be able to be there, but I have some other business to attend to."

"When could we begin?"

"I think Officer Davis is available. Just a moment." Johnson dialed a number.

"Officer Davis? Ryan Hughes and his party are here if you're available. Yes. Thanks." Johnson hung up the phone. "He'll be right here. Dr. Atkins, I understand that you helped develop the software."

Allen seemed surprised that he knew that. He looked at Ann who was looking away. "That's correct. The underlying software was developed by a colleague. I just came up with the application. We have spent quite some time putting data from solved cases in it. This will be the first time it is actually used in a real case."

Johnson scribbled in a notebook before he looked up at Ryan. "And what is the point of the interview today? It seemed unusually urgent."

"The trial date is approaching, and we just want to help Ross remember the events of the party more clearly."

"I guess there is no harm in letting you do that. I would like a brief synopsis of your interview, especially if anything comes to light that might implicate someone else."

"Of course."

A knock on the door interrupted him and Officer Davis entered.

"Well... here is Officer Davis."

Officer John Davis was a grizzled veteran, a no-bullshit type of guy.

"Officer Davis, I'll let everyone introduce themselves. I must go to court today on another case. I would appreciate it if you would escort them to Room 5 and monitor their interview of Ross Daniels."

Davis was as stiff as a board. "Yes, Sir."

"Thanks for your co-operation," replied Ryan.

Johnson shook their hands and each person introduced themselves to Officer Davis. "Well, shall we go?"

Phil and Allen picked up the suitcases and followed Ann and Officer Davis out of the office.

CHAPTER 13

The equipment had been set up and Phil and Ann were helping Allen make some last-minute adjustments. In the repackaging of the system, Allen had added several additional sensors to help Sherlock gather data. The usual array of windows popped open on the laptop screen, indicating that everything was functioning normally. Sherlock had already identified Ann and Allen and noted that there were four other "subjects" in the room. Adam was talking to Ryan when Phil signaled to Officer Davis that they were ready. He nodded and left to get Ross.

01:34PM ONE UNKNOWN SUBJECT LEFT THE ROOM. THREE UNKNOWN SUBJECTS IN THE ROOM.

"Ryan why don't you sit there, and we'll check it out on you."

"All right."

01:35PM PHIL CONLEY IDENTIFIED VIA VOICEPRINT, TWO UNKNOWN SUBJECTS IN ROOM.

Ryan sat down, and Allen adjusted the suitcase containing the infrared camera and laser, while watching the screen on the laptop. Labels on the appropriate windows opened and changed from Inactive to active.

"Thanks, that's it."

Adam was still looking at the screen when Ross Daniels entered with Officer Davis.

01:40PM TWO UNKNOWN SUBJECTS ENTERED ROOM. FOUR UNKNOWN SUBJECTS.

Ross Davis had just graduated from Stanford with a degree in Business Administration. Ross seemed to exemplify the Southern California lifestyle, with his muscular build and close-cropped blond hair. Adam quickly hugged his son.

"How are you?"

"Okay, I guess."

"This is Ann Stevens, my old friend, Phil Conley, and Allen Atkins. My son Ross."

They all shook hands.

"It's been a long time, Phil." Ross noticed the equipment on the table. "Is this it, dad? The system you were telling me about?"

"This is it. And these are the experts that I hope can use it to set you free."

Ross seemed eager to look at everything, but he stopped and glanced at Adam.

"It's great. But why did you bring it here?"

"We'd just like to ask you some questions about the party and what you can remember before you went to sleep."

"But you could just ask me. You don't need this. I thought this was for the witnesses who claim they saw me leave Bill's room that night?"

"It's for that too," replied Allen. "We would like to get some additional information for a new software system that helps solve crimes."

"That's part of this too?"

"Yes. In fact, I think it's time to introduce Sherlock." He turned the laptop speakers on.

"Sherlock?"

A new, improved Sherlock replied with a booming voice that startled Ross and Adam. "Yes, Allen?"

"I would like to introduce you to Ross Daniels and Adam Daniels."

"Please to meet you both."

Adam and Ross glanced at each other with shocked expressions.

Ann laughed at Sherlock's new and distinctly New England accent and their expressions.

Phil motioned Ross to sit down. "Why don't you have a seat and we can begin?"

Ross was still a little dazed but he managed a weak, "All right."

"Just act like he's not here," suggested Allen as they stared at the laptop.

Adam sat down next to Ross, and Sherlock boomed a question at them. "Who is seated in the interrogation chair?"

"Ross is." Allen laughed as Ross and Adam stared at each other. "Ross, tell us about the party again. Everything that you can remember."

"But I've been over this a million times with the police."

"I know, but with this system, we may come up with something that leads to finding the person who really did kill George."

"We are here to help you Ross," Phil reminded him.

"I know that."

"Would you rather that we asked you some questions?"

"It doesn't matter, either way. Some friends and I were invited to a party at Bill Kennedy's house. He's a friend of mine from college. My girlfriend Jill and I got there about 9:00PM and the party had been going on for a little while. Everyone had been drinking a lot. Some people had even jumped into the pool with their clothes on."

Phil had been making notes but glanced up to see Allen looking at him and pulling on his ear.

"Ross, were you actually invited to the party? Or was it like a lot of parties, where you and your friends heard about it and just showed up?"

Ross looked surprised. "Well, I guess we weren't actually invited, but Bill shows up at all of my parties."

"Why didn't you tell us this before?" Adam's voice betrayed his concern that Ross had not told them everything.

"It didn't seem important, whether I was actually invited or not."

Phil looked up from his legal pad where he had been busily marking up Ross' first statement to the police. "Knowing exactly what happened is very important if we are to help you. You also said that you saw everyone drinking a lot. Did you see anyone using drugs?

"No."

Allen was pulling on his ear and Phil decided to restate the question.

"Would it be fair to say that you don't know if anyone there was using drugs or not, only that you didn't see them doing it."

Ross pondered that for a moment. "That's right, I didn't see anyone doing it."

"Is drug use common at these parties?"

"I wouldn't say common, but it happens sometimes."

"So, what were you drinking?

"Just beer. I have a weakness for certain types of beers from micro-breweries, and Bill had a big selection of them, so I guess I got carried away and had too many."

"What happened then?"

"I knew I had too much to drink, and I didn't want to drive home, so I went upstairs and found a bedroom with no one in it, and I guess I passed out on a bed."

Ann was reading the same police report and wondered why Ross couldn't find a ride home. "Where was Jill when this was going on? Couldn't she drive you home."

"She's a best friend of Sheila, Bill's girl friend. They left pretty early to pick another friend of Sheila's from the airport."

Phil furiously scribbled some notes on a copy of Ross' deposition. "Did anyone see you go upstairs and into the bedroom?"

"A girl I met at the party helped me up the stairs and into one of the bedrooms."

"What was her name?"

"I don't know. I had never met her before, and I don't think she told me her name."

"Surely someone else saw you go into that bedroom."

"I guess some people did. There were a lot of people there, but Bill's parents have a huge house." He looked at Adam for a moment. "It's even bigger than ours. And there must be a dozen bedrooms upstairs. Someone may have seen me go into that one, but I don't remember seeing anyone else up there."

Allen noticed Ross' profile when he turned to look at Adam and frowned. Something seemed familiar about Ross, but he just couldn't place it.

"What about your friends? Did any of them see you go upstairs?"

"No, they were too busy trying to pick up some models that had been invited to the party."

"What about Bill? Where was he when you arrived?"

"He was trying to pick up one of the models too."

"What about the victim, George Artoles?"

"I really didn't know him very well. I would run into him at a party every now and then, but never really knew him."

Phil thumbed though some documents and pulled out a deposition given by a friend of Ross. "One of your friends said that George crashed your last party and that the two of you got into a fight and threatened each other."

"That wasn't anything. He was trying to sell some coke to some friends of mine at the party and I told him to leave."

"Several witnesses said the two of you started fighting and had to be pulled apart."

"He was just being a jerk. I told him I didn't want anyone doing drugs in my home."

Adam had a pleased expression on his face.

"They also said he threatened to kill you."

"He had a big mouth, that's all."

"They also said you threatened him back."

"I was angry, that's all. I didn't mean anything. I couldn't kill anyone."

"The prosecutor will be using this in their case against you."

Ross shrugged. "It's all bullshit. I didn't kill him."

"Did you see George at Bill's party?"

"I don't remember. There were a lot of people there. I know I didn't talk to him."

Allen rubbed his nose as Phil continued. "Ross, are you sure?"

"Yes, I'm sure."

Ann saw Allen rub his nose again. She could see the concern building on Phil's face.

"Ross, think carefully, did anyone ever mention that George had a drug problem or anything that might be related to that, like a gambling problem, or a drinking problem."

"I heard at one party that he was having trouble with his family, but the people who told me that were sort of enemies of his, so I didn't pay much attention."

"Do you know anything about his family?"

"Just rumors that they are into drugs and other illegal things. Bill said they are pretty big in the drug-smuggling business, but I never really had anything to do with them."

"You never witnessed any drug transaction involving George or a member of his family?"

"I've never met any of his family."

Allen was rubbing his nose again. Phil was trying to keep his anger from showing. Why was Ross not telling the whole truth? He could see that Allen was becoming frustrated as well. He decided not to confront Ross yet. The system could be wrong, and it was hard to believe that someone accused of murder would be lying to the team trying to help him. Unless he was covering something up.

"Why do think there are five people willing to testify that they saw you come out of the bedroom where George was found and run down the hall?"

"I just don't know. It couldn't have been me. I was out of it on the bed when they said he was murdered. They must be mistaken."

"You know that they all picked you out of a lineup."

"Yes. I just don't understand it. I didn't kill him. Why would I?"

"What about the one hundred thousand in cash they found in your car."

"I don't have a clue about that. It's not mine."

Phil strained to keep from yelling at Ross when he saw Allen rub his nose again.

"The State's case is based on the assumption that you were trying to make some sort of a deal with him, and it went badly. They also have the witnesses, which with other evidence, like the victim's blood on your shirt, will probably be enough to convict you."

"It's all bullshit." Ross alternately looked at Allen, Phil, and Ann. "So, can you help me?"

"I hope so. Ryan, is there any chance they would let us interview one or more of the witnesses? Perhaps even videotape their responses."

"I seriously doubt it. They have these people under tight security. We can ask, though."

"Do you know if they made any videotapes of their statements? And, if they did, could we look at it?"

"That isn't usually done either, but I'll find out. We have a right to see every bit of evidence against Ross. A videotape wouldn't be much better than a transcript for this system, would it Allen?"

"Not quite, but there are certain aspects of it that might be useful."

"I'll certainly ask, then."

"Well, I don't have any more questions for right now. I think this has been useful though. Adam, we can leave if you would like to visit with Ross for a while."

"Yes, if you don't mind."

Phil, Ann, Allen, and Ryan left them alone.

CHAPTER 14

Adam was seated as Phil, Ann, and Allen entered.

"Ryan is checking on the videotapes. So, what did you learn from this? What are his chances?"

They all sat down at the table and there was an awkward silence that Allen finally broke. "The good news is that I think he is innocent. The bad news is that I think he is covering something up."

Adam was shocked. "What do you mean?"

"Some of his answers were borderline lies."

"Are you sure?"

"I can't be one hundred percent sure, but there was a clear undertone of falseness in some of his answers."

Adam was distraught. "I just can't believe he would lie about this. He knows the seriousness of the charges against him. What could possibly be more important than the outcome of this trial. His life is on the line and...."

Ann interrupted him. "Should we bring him back here and confront him with this?"

"That would only raise suspicions about his denials and his plea of not guilty."

Ryan entered with a pleased expression. "Good news. A local TV news crew interviewed some of the witnesses the night of the murder. They have several copies here, and we can borrow one for as long as we need it."

Adam was delighted at the news. "That's great!"

Phil's stomach growled menacingly. "Maybe we could analyze the interviews after lunch."

"Oh, I'm sorry. I'll get Jacob over here immediately to take us somewhere. Excuse me."

As he left, Ann saw Allen staring out the window. "What's wrong, Allen?"

"I don't know. Something about the interview with Ross. I'm sure I'll think of it."

"We did pick up a few tidbits for Sherlock."

"Sherlock! I forgot all about that. It looks like I'll have plenty to do this weekend entering all that data."

"Can I help? Or could we get someone in to do the data entry for you?"

"Unfortunately, the interface to make that easy isn't finished yet. Most of the data must be coded before it can be entered, and it would take too long to teach someone how to do that. Phil knows how now, so we'll just have to take turns entering data."

Adam entered. "Jacob's here. Are we ready to go?"

They all indicated they were and left.

Ryan pressed the PLAY button on a digital video recorder. The infrared video camera and microphone were pointing at the TV screen. The blinds were closed, and Adam, Ryan, Ann, Phil, and Allen were watching the interviews of the witnesses. It was just about finished. Allen glanced at the monitor every now and then.

"...and then we heard a scream from Marilyn. I guess she had just seen the body. The police ran up the stairs and right by me to the bedroom. A few minutes later they came out and started asking everyone if they had seen anyone leave the room, and we told them we saw this guy run down the hall. They started looking in the bedrooms and found him asleep on a bed. Can you imagine that? He just killed someone and then fell asleep on a bed!"

The TV reporter seemed just as surprised. "He's just a suspect at this point, isn't he?"

The eyewitness shrugged his shoulders. "I don't know, I saw some blood on his shirt when they brought him down the stairs."

"Mr. Haynes, we really appreciate your taking time to speak with us tonight. I know all this must be upsetting."

"No problem. Hey, is this live?"

"Yes, but...."

Haynes started waving at the camera. "Hi, mom!"

The video ended, and Ryan stood up, turned the TV off and pressed the STOP button on the DVR. Ann stood up and opened the blinds. They were all looking at Allen.

"Based on the infrared imaging and voice stress, three of them are telling the truth and two of them are lying about what they saw."

"How can that be true?" asked Phil incredulously. "They all told the same story?"

"I don't know. It doesn't make sense."

Adam looked down, hope fading for a quick end to the family's ordeal. Ryan turned off the DVR and sat down.

Ann empathized with Adam and Ross' predicament. "Now what, Phil?"

"I don't know. Maybe Sherlock can help us when all this data is put in. Ryan, did you say the trial starts next Wednesday?"

"Yes, at 9:00AM."

Allen had been trying to rationalize the information they had gathered that day. He glanced at Ryan. "I would like to visit the Kennedy house just in case there is something missing from the enormous file Phil gave me. Could we do that on Monday?"

"I'll check with Detective Smith, but I'm sure it's okay."

Phil thumbed through his day planner. "Tomorrow's Thursday. Ryan, I would like to spend the whole day with you going over all the evidence the State has and try to refine the defense strategy."

"Sure thing. We can use my office. I'll arrange for some clerical and paralegal help for us."

Allen finished stuffing his briefcase with case files. "I'll bring Sherlock along and fill in some gaps in the database we have so far."

"Phil?" Ann stood up. "I don't think I can help much with the legal stuff. Several financiers want to talk about a possible investment in my company tomorrow, so I guess I'll see you Monday." She waved goodbye as she left.

Adam seemed somewhat dejected. "Allen, please let me know if there is anything I can do to help you. I'm feeling kind of useless right now, and I want to help."

"I promise I will, but right now, I need to take this stuff apart."

He started to disassemble the system and the others gathered around to help.

CHAPTER 15

Phil closed his laptop, sat back in his chair, and stretched. He stood up and walked out onto the balcony to watch the last rays of the sun as it set behind some dark, ominous rain clouds. He had just finished answering his email and was about to call for room service when the doorbell rang. He was surprised to find Ann holding a large bottle of champagne and two glasses. She had been drinking heavily and was leaning against the doorframe.

"Can I come in?" she asked unevenly.

"Ann, are you okay?"

"Sure. I just wanted to share my little celebration with you."

He took her hand and gently pulled her inside and closed the door. "What are we celebrating?"

She handed him a glass and filled it with champagne. "My divorce became final today." She filled her own glass and emptied it quickly.

Phil was starting to become concerned. He had never seen her this drunk. Ann usually had an enormous capacity for alcohol.

"I'm sorry it finally came to that; I know you and Dan really love each other."

She laughed weakly. "You're right. We just can't stand each other enough to live together."

She started to fill her glass again and Phil stopped her. "Ann, drinking won't solve anything."

"No. But it helps me forget." The champagne bottle slipped out of her hand and landed on her foot. "SHIT!" She grabbed her foot and almost fell down. Phil quickly grabbed her arm and helped her sit down at the desk. He took her shoe off and rubbed her foot. Her eyes were red, and her makeup was smeared from crying.

"You'll be fine tomorrow."

"No, I won't." She stared at him for a moment. "Phil?"

"Yes?"

"This has been a particularly bad day for me. Would you keep me company tonight."

Phil started to protest, but she held up her hand. "I just need a friend, that's all. I know you and Joanne have a thing and I'm not going to try and mess that up."

He knew how hard the divorce must have hurt her. Ann could be tough as nails in business dealings, but she was very vulnerable in her relationships. She was a mess emotionally as well as outwardly. He wanted to say no, but she started crying softly. She stood up to leave and he stood up and put his arms around her. She buried her face in his shoulder as she tried not to cry but couldn't. Her perfume brought back a lot of memories he had tried to bury. Ann had a passion for a particular French perfume that was hard to find, and Phil had once spent the better part of a day trying to find it. He still liked it on her. When she finally stopped crying, he held her back a little and looked into her eyes.

"All right. You can stay."

She smiled at him through her tears.

He quickly clarified his invitation. "As a friend."

The drinking had taken its toll and after a plentiful meal in the room, Ann had fallen asleep quickly. Phil was having a hard time sleeping and had watched television for awhile before he yawned. Finally! He checked on Ann and stood looking at her as she slept. He hated to admit it, but he still had some pretty strong feelings for her. Ann had fallen asleep in her underwear and had tossed and turned in the bed throwing the covers back. He remembered watching her sleep like that years ago. He fought off several lustful thoughts, pulled the covers over her and went to the bathroom.

Joanne looked at her watch. It was only a little after 11:00PM in Los Angeles. She knew Phil was a «night owl» of sorts and would probably still be up. She had some news she wanted to share with him and had finally found enough nerve to tell him. She took out her cell phone and quickly dialed his number.

The room phone rang several times before Ann reached out groggily and picked up the receiver. "Hello?" There was no reply and she hung up the receiver as Phil rushed out of the bathroom holding a toothbrush.

"Who was that?" he demanded. She didn't answer and he shook her. "Ann, who was on the phone?"

Her eyes flickered open for a second. "They didn't say anything. It was probably a wrong number." She closed her eyes and rolled over, facing away from him. Phil wondered if it were Joanne. Maybe he should call her? He glanced at the clock on the nightstand. One AM in Houston... it couldn't be Joanne. She was an early riser and went to bed early. He was hoping it wasn't her as he walked back to the bathroom.

Joanne stood frozen with the phone in her hand. At first, she thought she had dialed the wrong number. But the voice on the other end sounded familiar. Who was it? It suddenly came to her... Ann Stevens. What was she doing in Phil's room... at one in the morning? She suddenly felt drained of energy and fell back onto the bed.

Maybe she should call again? She was alternately feeling angry and betrayed. She took several deep breaths and forced herself to calm down. She would call him tomorrow and have it out. She realized she was still holding her phone and threw it on the bed.

Nature's call woke Ann and she walked quickly to the bathroom. The clock on the nightstand showed 4:00AM as she returned to bed. Phil was sleeping soundly as she crawled in next to him. His cologne was faint, but it brought back some fond memories she knew she would never forget. She was reminiscing about some of her favorite times with Phil when she realized she was playing with the hair on his chest. She softly stroked his face and smiled when he took a deep breath and sighed. She knew Phil slept like a rock. Phil suffered from a mild case of insomnia, and once he did fall asleep, he had difficulty waking up. He often struggled to be at work by nine in the morning.

Ann felt only a little guilty as she took her underwear off and started kissing him. She leaned over and started to lick his nose and lips. She didn't know why, but that always seemed to arouse him while he was sleeping. She laughed softly as she rolled over on top of him.

"It's too bad you won't remember this," she whispered in his ear.

Joanne woke up feeling depressed. She decided to phone Linda, her best friend and ask for her advice.

"Maybe it's nothing. Why don't you call him?"

"I can't... during the day. I never know where he'll be."

"Does he have a cell phone?"

"Yes, but he never has it on, he hates being interrupted."

"Call him tonight, then."

"I won't call him... the bastard!" she shouted at no one.

"Joanne, do you love him?"

She reluctantly replied, "Yes."

"Then don't give up on him. Find out what happened."

"I'm not sure I could take the truth if they were together."

"What if they weren't? He doesn't know, does he?"

"No."

"You have to tell him!"

"I can't now."

"Call him, Joanne."

She paused thinking, when Linda prodded her. "What have you got to lose?"

CHAPTER 16

Ryan's office was covered in paper: depositions, police reports, lab reports, almost every imaginable legal document associated with a trial. The defense team had spent the entire morning poring over the evidence, hoping to find some tactic that would convince a jury of Ross' innocence. Ryan had dismissed the other team members when Phil seemed satisfied, they had a detailed summary of all the admissible evidence in the case.

Phil was pacing the floor studying a synopsis of the case prepared by Ryan's assistant attorney and his paralegal.

"Let me summarize the State's case. Ryan, stop me if I stray from the facts and jump to conclusions." He glanced at Allen.

"I'll verify the database as you proceed. Ready when you are."

"Bill Kennedy threw a lavish 4th of July party and invited a lot of friends and people in the entertainment field. That's not surprising, since his father owns newspapers, TV and radio stations. Bill didn't specifically invite Ross to the party, but that's not unusual either, since they are good friends and go to a lot of parties together. That would also explain why Ross' name doesn't appear on the list of people Bill invited."

"Probably a hundred or more guests showed up at various times, but there were 54 guests there when George Artoles was murdered. George wasn't really a friend of Kennedy but was a source of illegal drugs on occasion. He initially showed up alone but left and came back with Mark Dunlop. It appears that Dunlop is a high-volume dealer and George was just a go-between with Kennedy."

"Dunlop, George, Bill, and an associate of Bill's, Sam Rockey, went upstairs around midnight for no apparent reason. The State believes this is when there was an attempt to make a deal for a large amount of cocaine, probably enough for everyone at the party."

"Midnight also happens to be about the time two patrol cars arrived in response to a noise complaint from a neighbor. One drove up to the front of the mansion while the second blocked the gate to prevent anyone from leaving or entering. The officers had just entered the house and were beginning to question some of the guests to find the host, when there was a gunshot, and someone screamed. The officers ran up the stairs and several guests pointed to an open door in the hallway to the right of the stairs. The officers entered with their guns drawn and found the victim on the floor bleeding profusely from a gunshot wound in the chest. The gun was lying next to him on the floor and there were three persons present in addition to the victim, Kennedy, Dunlop and Rockey."

"The officers yelled for them to not move but, according to the officer in charge's notes, Sam Rockey yelled back at them that the murderer had just left the room. One officer remained in the room and called for an ambulance while the other left to look for the possible murderer. He asked several guests who were standing near the stairs if they had seen anyone leave the room and they said they saw someone run down the hall past them. The officer began to search the rooms to the left of the stairs. In the last room he found Ross Daniels lying on the bed, apparently asleep. He tried shaking him, but he didn't respond. He also observed that Ross' shirt had a large bloodstain in the front. He handcuffed him to the bed and went back to report to the officer in charge. At this point, Ross is merely one of several suspects. And... the blood on his shirt is the only physical evidence linking him to the murder."

Phil paused to look at Ryan and Allen. Ryan gave him a thumbs-up sign. "Great summary. Please continue."

Allen was checking the data in the database against Phil's summary and didn't answer, so Phil continued.

"George Artoles died before the ambulance arrived. The officers questioned Kennedy, Dunlop and Rockey extensively, and indeed, Kennedy had been trying to buy enough cocaine for the party. According to Rockey, he was there to help Kennedy make the deal. Rockey stated that George had recently sold some low-quality cocaine to Kennedy at a premium price, and Bill wanted him there to make sure he didn't get cheated again. Rockey brought a test kit that could check the cocaine for purity and potentially harmful additives."

"According to Rockey, Artoles must have given Kennedy some cocaine earlier because he was barely coherent, and he had to finish negotiating the price with Dunlop. Rockey was about to test the cocaine Dunlop had brought when Ross Daniels entered, looking for Kennedy. Neither Dunlop or Rockey knew Ross, but they said he tried to talk Kennedy out of buying the cocaine but was interrupted by George. Ross and George began a loud argument. They were pushing each other when George pulled out a pistol. Ross also had a gun in his jacket and managed to shoot George first, hitting him in the chest.

"George fell forward onto Ross who pushed him off and fled through the door. Dunlop and Rockey were trying to stop George's bleeding when the officers arrived. Kennedy had flushed the cocaine down the toilet and passed out in the bathroom. There was no evidence of drugs, other than the fact that Kennedy had obtained some cocaine earlier from George and had a significant amount in his blood."

"During the initial investigation, a bag containing $100,000 was found in Ross' car. The State assumed Ross was somehow involved in the deal and had brought the money to the party."

Phil slapped his notepad down on Ryan's desk. "This is bullshit! Why would Ross bring money to a party so that Bill could buy drugs? Ross had thrown George out of his house for trying to sell drugs to his friends. They had even had a fight over it, and now the State is trying to use that fight against Ross as evidence of bad blood between them and...."

Ryan interrupted him. "The State knows that this is a weak link in their case. Mike told me this bothered him. He doesn't have a good explanation of why Ross would want to help Bill buy drugs. But they have all the other evidence, including the blood on Ross' shirt and the witnesses who saw him leave the room where George was murdered."

"What if Ross were bringing the money for a different reason and didn't even know about the drug deal. It might just be a coincidence," suggested Allen.

"Where would Ross get that kind of money? And he says he knows nothing about it. He appeared to be telling the truth about that, at least." Phil's anxiety was beginning to show.

"Why don't we come back to this later?" suggested Allen.

Phil nodded. "You're right." He picked his notepad back up and resumed his summary. "Rockey, Dunlop and Kennedy were arrested as well on various drug-related charges, but the State had no evidence at the scene of the crime and most of the charges were later dropped. Rockey, Dunlop, and Kennedy served short sentences in the county jail and were released. Ross was charged with second degree murder and denied bail."

"The police took depositions from several witnesses, who said they saw Ross leave the room after the gunshot and run past them down the hall. In separate questioning, Dunlop's story was identical to Rockey's. Kennedy had been too stoned to remember much, other than Ross and George's fight and the gunshot. He refused to answer any questions about the cocaine or the source of the money that was found in Ross' car."

Phil sat down wearily and tossed his legal pad onto Ryan's desk.

Allen shook his head in disbelief. "Quite an incredible story, and the big loser appears to be Ross, the one least likely to be involved in something like this."

Phil rubbed his eyes. "Ryan, is there any hope of a postponement? Even a few weeks would help."

Ryan shook his head. "Judge Burns has denied all requests for a delay so far. Once he makes up his mind, he rarely changes it."

Allen had entered a massive amount of data into Sherlock's database. Ryan and Phil waited for him to finish typing. He finally looked at them.

"It's running."

"Let's get a quick cup of coffee while Sherlock digests the latest data," suggested Phil.

Allen plugged a set of speakers into the laptop and they left to stretch and relax for a moment. When they returned, Phil observed that the laptop's speakers were off.

"Why don't you just leave the speakers on, Allen?"

Allen shook his head. "When I first received the latest version of Sherlock from Tom, it seemed that it was spending as much time trying to find out what I thought about the data as it did analyzing it. I thought Tom may have sent the wrong thing, so I called him to ask about it. He checked with his wife and found out they accidentally included a beta test module known as 'Sigmund Freud'."

Phil laughed. "Sigmund Freud?"

"Yes, it has all the basic elements of Sherlock Holmes' scientific approach to solving crimes but adds a psychological approach as well. I told him I didn't want that. Unfortunately, it's a more advanced version and all the data I had entered since I received it was converted to the new module's format. If he gave me the latest version of Sherlock without the Freud module, it wouldn't be able to read it. That would include all the data we collected in the interviews at Ann's business."

"That would pretty much render the whole thing useless, wouldn't it?"

Allen nodded. "It looks like we're stuck with the Freud version, whether we like it or not."

Phil glanced at the laptop screen. "Allen?" was flashing on it. "Is there any downside to the new version, other than the fact it wants to know what you think?"

"No, it's just damn annoying. That's why I keep the speakers turned off."

Allen turned the speakers on and folded his arms in frustration.

"Sherlock?"

"Yes, Allen?" Replied the deep booming voice.

"Have you completed an analysis of the latest data."

"Yes."

"Are there any inconsistencies?"

"A few minor ones that I would like to talk to you about."

Allen sighed. That was not good news. The start of the trial was imminent and left little time to gather any more information.

"I guess that means the State has a pretty strong case?"

"Yes... based on the circumstantial evidence."

When Phil returned to his hotel that evening, the message light on the hotel phone was blinking. Joanne had left a message to call her as soon as possible. He dialed her home number.

"Hi, Joanne, it's Phil."

She didn't immediately answer and he wondered if something were wrong.

"Phil?"

"Yes?"

"I wanted to discuss something with you, and when I called you late last night, a woman answered your phone."

Damn, it was her. He hoped she would believe him.

Joanne had not spent the day idly wondering what had happened. She drove to the NASA Space Center and met Janet Turner. Janet had buffaloed Ron into believing that Allen needed his modified Voice Stress Analyzer, and that Joanne was going out to visit Phil and would take it with her. He had gladly handed it over to Joanne who was now holding it against the receiver.

Phil relayed an incredible tale of why Ann was in his room and had happened to answer the phone. He also stressed that nothing had happened between them. When he finished, he waited for her to say something. Joanne was still staring at the VSA, which had not blinked. Her mother used to tell her that trust was a cornerstone of any relationship and that once it is broken, it may never be restored. She suddenly felt guilty at using a lie detector on Phil and tossed it onto the sofa next to her.

"That's a pretty incredible tale, Phil."

"I know... but it's the truth."

"I believe you, but...."

"Don't let it happen again?" he volunteered.

"You got it."

Phil felt a great sense of relief. He had dodged a pretty serious bullet and was determined he wouldn't be put in that situation again.

"How's the trial stuff going?"

"Not very well. We still haven't found a good defense strategy yet."

"Is Sherlock helping?"

"Some. We're going to visit the murder scene on Monday. I'm hopeful we'll find something there. By the way, why did you call?"

"I saw something on the news here, that the trial was about to start," she lied, suddenly afraid to tell him the truth. "I guess you'll have a pretty busy weekend...."

"No, I'll just be helping Allen enter more information into Sherlock's database. We take turns typing."

"Call me when you aren't too busy."

"You know I will... I love you, Joanne." He tossed that out, hoping she wasn't still mad at him.

"I love you too, Phil."

When Joanne hung up the phone, she took out her checkbook and reviewed her financial situation briefly. She picked up the receiver and dialed a number. After a brief wait, an agent answered "Sun Travel Agency. Can I help you?"

"Yes, I need some information. What is the cheapest flight to Los Angeles tomorrow afternoon?"

CHAPTER 17

Allen was intently typing on the laptop when the hotel phone rang. He absently pressed the speakerphone button. "Hello."

"Hi, Allen, this is Alicia. Dad said that you were spending the weekend putting data into Sherlock, and I was wondering if you would like to take a break and have dinner with me."

He looked at his watch. "Sure, I'll catch a taxi and be over in a little while if that's okay."

"Oh, you don't need to do that, I'll pick you up."

The doorbell rang, and Allen stood up. "Alicia, can you excuse me for a minute, there's someone at the door."

"Sure."

He put the phone down, and opened the door. Alicia was standing there in a business suit with a cell phone to her ear. "Are you ready to go?"

Allen smiled as Alicia entered, shutting the door behind her. "Give me a minute to clean up."

"Take your time. I'm not in a hurry."

Allen replaced the receiver on the phone and pressed a few keys on the laptop, shutting it off. Alicia opened the patio door and walked out, shutting the door behind her.

A few minutes later, Allen walked out and stood next to her. "Where would you like to eat?"

She suddenly put her arms around him and kissed him hard. "We could order room service and spend the evening here."

Allen fought off a wave of desire and managed a weak reply. "I think we'd better go out."

"Okay, Allen," she said wistfully, as she walked inside and picked up her purse. Allen realized he was almost panting.

Allen and Alicia were soon seated at a corner booth. The waiter left, writing their order on a pad.

"So how's the case going?"

"Pretty good. I hope to finish entering this week's data into Sherlock by tomorrow, then we'll get together to review it. How did you know I like Mexican food?"

"I have my sources."

"Probably Phil."

She smiled and then gazed deeply into Allen's eyes. He began to feel a little uncomfortable, and looked away for a moment.

"You have pretty eyes."

"Thanks, Allen."

"Alicia, you could have asked about the case and Sherlock on the phone."

"But that wouldn't have been any fun."

"Having dinner with me is fun?"

She stared into his eyes. "It could lead to fun."

Allen felt oddly embarrassed yet excited. He was still in awe of her cover-girl looks and quick wit. "I should have told you that I have a girl friend back in Houston."

"I'm not interested in her."

Allen was suddenly confused. "No, I mean...."

"I know what you mean. I have a boy friend too."

Allen breathed a little easier. "So why aren't you having dinner with him? Is he out of town?" Allen hoped she would say yes.

"No. I'm attracted to YOU, Allen."

Allen realized he was sweating. "You don't really know me."

"I know more about you than you realize."

Allen coughed nervously. "I'm going to have to have a talk with Phil. Are you always this direct?"

"Usually... I'll bet when you were in high school, there were some girls that you didn't ask out because you were afraid they would say no. Am I right?"

Allen thought about that for a moment. "Yes. More than a few times."

"What if you had asked and she had said yes?"

"It would have been great! But what if she had said no?"

"Ask someone else."

"I wish it were that easy."

"It is, Allen. You have to buy a ticket if you want to win the lottery."

Allen laughed. "Okay, Miss Direct. What is it that you want?"

"You, Allen!"

Allen felt himself blush. "What! What do you mean? A one-night stand?"

"No, but that would do for starters. Who knows where that might lead?"

"I told you; I have a girl friend."

She took a sip of her drink staring at him. "Are you engaged?"

"Well... no."

"Have you agreed to not date other people?"

"No."

"Are you thinking about getting engaged?"

"No."

"Are you even in love with her?"

Allen couldn't even remember telling Janet that he loved her. "I'm not sure."

"Well, then what's the problem? You wouldn't buy the first car that you test drove, would you?"

"No."

"You wouldn't buy the first house you looked at, would you?"

"No."

"Then, is there anything wrong with getting some comparison data?"

"I don't think so."

"That's the problem, Allen. You think too much."

Something in the back of Allen's mind suddenly bubbled out. "Well, what about your boyfriend?"

"He dates other women."

Allen laughed loudly. "He must be an idiot!"

Alicia chuckled at Allen's comment. She put her hand on his as the waiter arrived with a steaming plate of fajitas and some side dishes.

"Enjoy your meal!" he said as he gathered their salad plates and left.

Alicia held up her beer bottle in a salute. "To dessert."

Allen picked up his beer bottle and touched hers.

As Joanne exited the elevator on the 12th floor, she looked at the telephone number written on a small piece of paper. Room 12101 was at the end of the hall. Her resolve crumbled as she neared the door. This is a bad idea! She took a deep breath and forced her finger to push the doorbell button.

It had been a long week for Phil, and he was looking forward to a quiet weekend. He turned on the TV, kicked his shoes off and fell easily onto the bed. He picked up the room-service menu and was about to pick up the hotel telephone to order when the doorbell rang. He looked at his watch. Probably the maid to turn the bed down. When he opened the door, he dropped the menu. Joanne was holding a suitcase with a big, if somewhat embarrassed smile. Phil felt a sudden rush of energy followed quickly by an even bigger rush of desire.

"Joanne?"

"Hi, Phil. I was sort of missing you and thought you might like some...."

Before she could finish, Phil grabbed her and pulled her into the room, closing the door. She dropped the suitcase when he picked her up and hugged her tightly. He was kissing her and undressing her at the same time. When she finally could speak, she whispered in his ear. "I didn't come just for this."

Phil didn't seem to notice; he was too intent on making up for lost time. They both laughed when he pulled the bedclothes back so hard, everything flew onto the floor. Joanne barely had time to pull the covers back before she became the object of his desires.

When Allen and Alicia entered his hotel room, she pushed him aside and ran to the bathroom. Allen laughed. He had never known anyone quite like her. She always said whatever was on her mind. It sometimes embarrassed him, sometimes aroused him. He never knew what she would say or do next. He sat down on the bed wondering what would happen and began thinking about Janet and their last night together. He was feeling a little guilty until Alicia came out of the bathroom wearing one of his

T-shirts. When had she taken that out of the drawer? As she drew nearer, he realized she wasn't wearing anything else and he took a deep breath. She stood in front of him and pushed him back on the bed, crawling on top. They kissed deeply and she whispered in his ear.

"If you need to go to the bathroom, you better go now. You won't be getting up again for a while."

He rolled her off of him. He heard her chuckle as he hurried into the bathroom.

Joanne woke up and slid quietly out of bed so she wouldn›t wake Phil. The clock on the nightstand showed 3:00AM. She stood by the sliding glass door to the balcony, watching the city lights below twinkle in the dark and wondered how she was going to tell Phil why she really came out to see him. Visions of the little white stick turning blue kept flashing through her mind. Her stomach felt a little queasy. She wondered how he would react to her being pregnant. She would have to find the right time and place to tell him.

Phil suddenly put his arms around her, and she jumped so hard she almost hurt a back muscle.

"Sorry," he whispered in her ear.

He hugged her and pulled her gently to the bed. Even in the dim light of the room, he could see that she had been crying. He was dumbfounded. It had been wonderful for him. Maybe in his eagerness he had hurt her. "Are you all right, honey?"

She started to cry softly in his arms, confusing him even more. "What is it? You can tell me anything. You know that."

She cried even harder, and he got up to bring her a glass of water. When she had finished it, he put his arms around her and waited for her to stop crying. She eventually whispered to him.

"Phil?"

"Yes?"

"Don't hate me."

"What are you talking about?" he asked gently, hugging her even more.

"I'm pregnant."

He let go of her to turn the light on. She instinctively put her hands over her eyes, afraid of what she might see on his face. She also was still embarrassed to be naked in front of him with the light on. He pulled her hands away from her face.

"Joanne," he said softly.

She opened her eyes. He was sitting in front of her.

"That's the most wonderful thing I think I've ever heard."

Her eyes got bigger. What did he say?

He leaned forward and hugged her tightly kissing her hair, her neck, her cheek and finally her lips. Joanne put her arms around him, and they hugged each other for awhile.

He took her chin forcing her to look at him. "Will you marry me?"

That was the only question she hadn't prepared an answer for. She stared into his eyes until she found the answer she needed. "Yes."

"Could we do it today?"

She was in a daze. "What?"

"If you want a big church wedding, that's okay, but I would be just as happy if we could go to Las Vegas today."

Joanne was too stunned to answer, and Phil seemed intent on starting the honeymoon at that moment. She was too happy to resist him. She laughed and yelled "Earthquake!" When he finally took a break, she whispered in his ear. "Phil?"

"What?"

"Could we do both."

"Both?"

"Go to Las Vegas today AND have a big church wedding?"

Why not? "Of course."

CHAPTER 18

Brian Limpanatti watched Phil and Joanne walk up to the registration desk. He slowly got up and wandered over near them to listen. They were dressed in very casual clothes, and both were carrying suitcases. He wondered where they were going. He pretended to read some literature at the end of the desk as he watched Phil give the clerk his key. As the clerk left to find his records, Joanne tugged on his sleeve.

"Phil?"

"Yes?"

"If we are going to spend the whole time in bed, then we might as well stay here."

Phil stroked her hair gently. "What do you have in mind?"

"I've never been to Las Vegas. I want to see the shows and go to a casino."

Brian's ears perked up. Did she say Las Vegas? They were coming into his domain. He would have to call Jake immediately. He walked nonchalantly back to the maintenance closet and pulled out his cell phone to call his boss. Much to Brian's disappointment, Jake insisted that he remain in Los Angeles and keep an eye on Allen. Jake had other men who could take care of Phil and his girl friend.

The plane arrived in Las Vegas in the early afternoon and Joanne was like a kid in a playground. She had to try every slot machine in the airport before Phil could finally get her into a taxi.

The first stop was a small wedding chapel where they bought the "Honeymooners Special" package that included simple golden rings, a wedding ceremony complete with organ music, and a wedding license.

118

Phil kept repeating his intention to replace her ring with a large diamond engagement ring. Joanne patiently replied each time that it didn't matter. After the ceremony, they checked into one of the newest hotels on the strip and Joanne almost dragged Phil into its casino. He waited patiently as she began playing the nickel slot machines. He started whistling to himself.

"Why don't you play?" she demanded, annoyed at his whistling.

"I don't like slot machines. Why don't you play the dollar slots or at least the quarter machines?"

"That's too much money."

Phil was smiling as he took her arm. "Come on. Let's go to the bar. There are a few things I need to talk to you about."

When they had settled into a comfortable booth in the lounge area and ordered drinks, Phil had all of Joanne's attention as he began.

"When I finished law school, I tried civil law first. It seemed relatively easy and it paid well. I joined a large, successful law firm. Most of their cases were settled out of court and the amount of money involved was the only issue. I first met Adam Daniels when he was being sued by several large shareholders of the large mutual fund he was managing. It was basically just a misunderstanding, and the case was settled out of court. Some of the back room deals that went on were a little hard for me to swallow though, and I began to think about moving into criminal law."

"Even though I was just a junior attorney in the case, Adam and I became friends. He even recommended me to several of his friends. I started to build up a pretty nice client list and a reputation in the firm. I was also making a lot more money than my lifestyle required. I saved a lot of it and put it into the stock market. Adam even offered some advice on how to invest it and it's done really well. I probably could retire now and have enough money to live on until I started getting social security... if I were careful of course."

Joanne was watching his eyes with such rapt attention that Phil wasn't sure she was even listening.

"Joanne?"

"Yes, Phil?"

"Are you listening?"

"To every word. Please go on."

"Eventually, I got tired of the dealings and double dealings and decided to try and make a difference. I first thought about being a public defender, but that didn't pay enough, and the caseloads were too high, so I applied to the District Attorney's office and eventually got this job as an assistant DA. The pay is okay, and the casework is not impossible, so I settled into a pretty routine lifestyle... until I met you, of course."

She smiled at him. "And now you're going to be a daddy."

He smiled and put his hand on hers. "And a husband."

"I'd like to hear everything about you, but why are you mentioning this now?"

"I guess I'm trying to tell you that while I'm not rich, you won't have to worry about having a good time while we are here, and you can still have the biggest wedding ceremony you want."

Joanne put her hands on his. "Money's nice, Phil. But I only need you to be happy... and our child of course."

Phil scooted around the booth to snuggle next to her. They started kissing passionately and several other lounge patrons noticed them and snickered.

Antonio Bracchus wasn't amused. He was waiting until he could get Phil and Joanne alone. His orders from Jake had been specific: Phil was not to return to Los Angeles... and it was to look like an accident. He signaled to the actors he had hired to begin.

Phil and Joanne were suddenly the object of a camera with several bright lights. A man with a bubbly personality and dressed in a loud, flashy suit shoved a microphone in Phil's face as a man holding a large video recorder maneuvered to get a close-up of them.

"Are you Phil Conley?"

Phil was reluctant to reply. He didn't like giving his name to strangers. "What do you want?"

"Hi! I'm Ralph Ohlmeyer with Resort Activities of Las Vegas. We have been hired by the hotel to greet and reward the hotel's one-millionth customer. If you are the Conleys, we have a lot of gifts for you."

Phil wondered briefly how this organization had identified them among all the people in the lounge and casino, but Joanne became excited. "Oh, Phil. Tell him who we are."

Against his better nature, he replied. "Yes, we're the Conleys."

"That's great, Mr. Conley. I hope you don't mind if we film this. The hotel would like to use this in its promotional activities."

"No... I guess not."

"That's great, then. Would you mind coming out into the lobby? We have some really great prizes for you and your wife."

"Oh, Phil! This is so exciting. I've never won anything before!" Joanne was bubbling over.

They followed Ralph into the lobby.

One of the «prizes» Phil and Joanne had won was a five-acre lot with «fantastic views» in a new housing development under construction in the nearby mountains. A glossy brochure proclaimed the lots ideal home sites. A site visit was required at which time they would personally select a lot, inspect it, and sign the deed at the development's office. They also received use of a sporty convertible rental car to get them there, along with some spending money and other nameless trinkets. It was a warm, cloudless afternoon as Phil pulled onto a highway leading out of Las Vegas. Joanne had a GPS map on her phone and was acting as navigator. The city was soon left behind and vegetation became sparse. Phil was wondering how a housing development so remotely located could draw significant numbers of potential clients. He had barely noticed that the roadway was somewhat elevated and that he hadn't passed a car in miles.

He glanced at Joanne. "Are you sure this is the right way?"

She stared intently at the GPS map for a moment. "I'm positive."

Phil saw two cars approaching in the distance, one apparently attempting to pass the other. Phil instinctively slowed down to give the passing car plenty of time to make it. He soon realized that the passing car was not making any progress on the other car. Both were coming right for them. He looked around quickly and couldn't spot a side road to turn off on and the shoulder of the road was too small to pull out of the way. A vision of them rolling over down the road embankment forced him into a quick decision. Joanne had been studying the map on her phone and not seen the approaching cars. Her head almost hit the dashboard when Phil jammed on the brakes.

"What is it, Phil?"

Phil jerked the shifter into 'Park', reached over and pulled on the inside door handle, pushed her door open and unhooked her seat belt.

"What...." she yelled when Phil pushed her out and jumped out after her.

Joanne rolled down the embankment with Phil right behind her. They were barely 20 feet away from the roadside when the passing car slammed into the rental car at a high speed. The gas tanks on both cars exploded sending a shower of burning gasoline and debris around them. The first car screeched to a halt down the road and then started to back up. Phil pushed Joanne's head down and managed to peek over the scrub brush in front of him. He watched as the driver of the other car stopped near the wreckage, got out and started looking around. Phil ducked down for a moment. When he looked up again, he saw the man talking on a cell phone. "This was no accident," he said to himself. He wondered if they could run away without being seen. When the man was looking in the other direction, Phil whispered to Joanne as he helped her up. They bent over and quickly walked several hundred feet to hide behind a clump of scrub brush and cacti.

Jorge Gomez had executed the attack on Phil and Joanne precisely as Antonio had ordered. He had connected a «sacrificial» car to his own with a small tow bar that broke off exactly as predicted on impact with the rental car. Jorge leaped from his car to find the victim's bodies and report back. He searched everywhere but couldn't find any trace of the man and woman who were supposed to be in the car. His dream of a big cash bonus was quickly fading away. His hands were shaking so hard he had a difficult time entering Antonio's number on his cell phone. Antonio would definitely not be happy.

A few minutes passed, and Phil and Joanne heard a helicopter in the distance. He quickly associated it with the accident. They were probably looking for them.

The wind rolled a sagebrush into him, giving him an idea. He told Joanne to get as close to a scrub brush as possible and he placed the sagebrush against her. The sun was starting to go down and he hoped they could hide out until nightfall. He pulled some more sagebrush around them and waited.

Antonio Bracchus was furious when he received word that Phil and Joanne were not in the wreckage. He had immediately dispatched a helicopter with two gunmen to find them and finish them off. He knew

Jake would not be happy if Phil and Joanne were found murdered in his home town, but he couldn›t let them get away now or Jake would take care of him.

The helicopter and the two gunmen circled over the crash site for over an hour but couldn't find a trace of Phil or Joanne. The highway patrol had found the accident and started an investigation, forcing them to move away from the scene. The police had first assumed the helicopter was from a radio station or a company in the area. They had tried to contact it without success and eventually called for a police helicopter to find out why it was there. It fled when the police helicopter arrived. Antonio had briefly considered shooting himself but decided to take his chances with Jake. He reluctantly dialed Jake's private number.

Phil recognized the police helicopter and led Joanne to the officers investigating the accident. At Phil's insistence, the police helicopter flew Joanne to a hospital to make sure that she and the baby were all right. Phil began a lengthy discussion with the highway patrol.

Allen had spent most of Saturday entering data into Sherlock's database. He typed in a «Run Analysis» command and picked up the phone to call room service. He had barely placed an order when the laptop beeped at him and a message from Sherlock appeared on the monitor. Allen only turned on the laptop speakers when he wanted to talk to Sherlock. Tom's latest version of the software seemed to reflect his wife's personality and either made comments out loud as the data was analyzed or wanted to talk to him constantly about the data. As a serious researcher, Allen was driven to distraction by Sherlock's constant chatter. He needed time alone to think about problems and possible solutions.

He turned on the laptop speakers.

"Yes, Sherlock?"

"Allen, there are several inconsistencies in this latest data that I would like to talk to you about."

Allen sighed. "Okay, what do you have?"

Michael Stone had spent several hours placing listening devices in Allen and Phil's rooms where they wouldn›t be discovered. His telephone taps employed the latest technology and all the data recorded was being stored on digital hard drives for later analysis. La Brea Security had even developed a sophisticated speech-recognition system that could instantly turn recorded conversations into transcripts with a high degree of accuracy. It saved Michael and his co-workers a significant amount of time when they reported the results of their latest eavesdropping.

Brian Limpanatti had managed to obtain the room next to Allen to set up a listening base for La Brea security and its highly sophisticated equipment. Michael Stone happened to be on the afternoon shift when Allen began his conversation with Sherlock.

Brian put his sandwich down and glanced at Michael who was listening intently to Allen's conversation with Sherlock. "Who's he talking to?"

"I don't know. I thought your men were watching his room. Didn't they see him come in?"

Brian picked up his walkie-talkie. "Hey! Guys! Who's in Atkins' room now?"

Mario had parked himself in the lobby, watching for anyone related to the trial to pass through. He replied first. "No one's come through the lobby, boss."

Juan was standing in the stairwell with the door propped a few inches open so that he could watch the hallway outside Allen and Phil's rooms. He closed the stairwell door and answered. "No one's gone to his room since room service delivered his lunch."

"Are you sure he's not on the phone?" Brian seemed annoyed at Stone. "I'm positive."

"Maybe a cell phone?"

"I'm monitoring all the normal cell frequencies. He's not on a phone."

"Then who in the hell's he talking to? How about a CB radio?"

Stone laughed. "I have that covered too, but Atkins doesn't seem like the type to talk on a CB."

Stone pointed to a piece of equipment on the rack in front of him. "See that, I've even recorded your conversations with your men on that private band radio you're using."

Brian appeared suddenly upset. "Hey! Don't record nothing of me or my guys," he said threateningly. "What other way could he be talking?"

Stone started counting them off. "Well, he could be using a private radio like yours, but I'm not picking anything up on those frequencies.

He could be on a ham radio or a short-wave radio, but we can probably discount those as well."

"Why?"

"They aren't making the usual call signs required for licensed communications."

Brian stared at him for a minute. "What?"

"They have to identify themselves each time they talk on those frequencies."

"Oh."

Stone stared at several indicators on the rack. "Whoever it is, they aren't on a radio anyway. The sound frequency is too broad."

Brian didn't even bother to ask him what that meant. He scratched his head as he stared at a copy of the transcript that was being printed even as the conversation continued.

"Who in the hell's Sherlock?"

Michael hadn't been paying attention and answered almost automatically.

"You mean the fictional detective created by Arthur Canon Doyle?"

Brian stared blankly at him. "Huh?"

Several pieces suddenly began to fit together for Stone. Based on some previous conversations he had recorded; it finally came to him that Allen might have created an interactive software program. Stone knew that police departments used special programs to help solve crimes. Maybe Atkins had one and had added a voice interface to it. It almost seemed too incredible.

Brian saw him shaking his head. "What?"

Michael relayed his idea, but Brian really didn't understand what he was talking about.

"Ryan seems like a pretty smart guy, maybe he can help." He picked up his cell phone, dialed Ryan's number and quickly passed it to Stone.

"Tell him your idea!"

Phil spent an uneasy night in the hospital. Joanne had been examined and then sedated to help her rest. The ER staff had tried to assure him that the baby was all right, but he wouldn't be satisfied until he talked to her. He was sitting in a chair beside her bed when she finally woke up.

When she saw his head on the bed, she smiled and tapped him lightly on the shoulder. Phil jumped up.

"What?" When he saw that she was awake and smiling at him, he gently hugged her.

When Joanne was released, Phil took her back to their hotel. They spent the rest of the Sunday afternoon snuggled in bed and caught an evening flight back to Los Angeles. Ann and Allen were overjoyed when they learned of Phil and Joanne's marriage and treated them to dinner at the finest restaurant they could find.

CHAPTER 19

Monday turned out to be a bright, beautiful day as Phil, Ryan and Allen pulled into the large circular driveway in front of an enormous mansion in Bel Air that almost seemed to dwarf Adam Daniels› huge estate. Ryan was carrying a large briefcase stuffed with trial depositions and police reports. He smiled at Phil's and Allen's expressions as they got out of his Lincoln. Phil and Allen couldn›t help gazing at the meticulously manicured landscaping and Greco-Roman statuary lining the driveway as they walked up to the front door of the Kennedy mansion.

"This is massive. What does Michael Kennedy do?"

"He's a media mogul. I think he owns several newspapers and a few TV and radio stations around the country." When Ryan arrived at the front door, he rang the doorbell and presented an ID to the maid and a security guard.

"Are you sure this is okay?" Allen glanced around the massive foyer. "Where are the Kennedys?"

"They're in Europe. They wanted to stay away during the trial. They have given the police permission to look around, and we have the same privileges."

The interior of the mansion was even more luxurious and ornate than Phil and Allen would have guessed. Comparisons to the castles and country estates of the wealthy and royalty of Europe came to mind as they followed Ryan through the foyer to a massive staircase leading to the upstairs bedrooms. "This way." He led them up the stairs to a large hallway lined with original oil paintings and statues. They stopped to look up and down the hallway.

Allen tried not to gawk. "This is even bigger than it looked in the videos."

Ryan led them down the hall to a bedroom and opened a door for them. "The body was found near the bed." He stepped aside for them to enter.

Phil slowly walked around the room examining the layout of the room and the exquisite late 18th century furniture. All evidence of the murder had been removed. Allen and Ryan stood by the door watching Phil open the bathroom and closet doors. After a little while he seemed satisfied. "Okay, let's go see where Ross was found."

They walked slowly down the hall, past the stairs to a bedroom at the other end and entered it. Phil repeated his surveillance of the room, eventually opening the door to a patio, and walking out. He looked down to the grounds and then up to the roof, eventually walking back into the room. Ryan and Allen were quietly standing by the bedroom door watching him. Phil sat down on the bed looking out the window. After a few moments of quiet meditation, they saw him smile.

"What is it, Phil?"

"Well, Allen, the only difference between this room and the other room is the patio."

Ryan looked confused. "So what?"

"The killer couldn't have escaped out a window where George died, but he could have here."

"What's your point?"

"Ross could have gone out the window and managed to climb down or jumped to the ground, and he would have gotten away. Yet he just fell down on the bed and went to sleep?"

"I know what you mean. It bothers me too." Allen made a few notes. "But what about George's blood on his shirt?"

"I don't know. I can't explain that or the one hundred thousand in cash in his car."

"I wish we could," replied Ryan thoughtfully.

"By the way, where was Ross' car parked that night?"

"He parked in the circular drive, near the house. I guess someone had just left when he arrived."

"Where? Can you show me?"

"Sure. Follow me." Ryan led them outside to an area near the house on the large circular drive. "The car was parked right here."

Allen and Ryan watched Phil walk up and down the drive, viewing the house at different angles.

"And the $100,000 in cash was on the floor of his car?"

Ryan opened his briefcase and consulted a summary of Ross' statements to the police and the police report by the officer in charge of the investigation. "Yes, right in front of the driver's seat."

"Ross had a new convertible, didn't he?"

"Yes, and the top was down that night. What're you thinking?"

"If I jumped out that bedroom window, I would have walked by here on the way out, wouldn't I?"

Allen quickly picked up on Phil's line of reasoning. "Right past the police that were here responding to the noise complaint."

Phil rubbed his chin, lost in thought for a moment. "And if I WERE carrying a lot of money on me, and the police were to stop me on the way out, they would see it."

"And I couldn't take the chance of being caught with it. So, I would have wanted to ditch it A.S.A.P."

"Very good, Allen."

"This is all speculation," observed Ryan.

"Of course it is. That's all we have at this point - reasonable alternatives."

Allen scribbled in his notebook. "That would explain the money, but not the blood on Ross' shirt.

Phil waited for Allen to finish some notes. "Maybe Sherlock can help us there."

"There's one other thing, Phil."

"What's that, Allen?"

"The name of the killer."

"I don't think Sherlock is going to give us that," Phil finished his notes. "I'm done here, Allen, unless you have something else you would like to see."

"No, I can't think of anything. But you have hit on some interesting possibilities."

Phil looked at his watch. "Ryan, we need to go. We have an appointment with Judge Burns at 11:00."

"Oh, you're right. I had forgotten about that."

Allen pulled his notepad from his suit coat pocket. "Ryan, could you drop me off at the police station? I would like to discuss some things with detective Smith."

"Sure. Let's go."

Allen and Phil compared notes as they walked to Ryan's Lincoln.

CHAPTER 20

Judge Harold Burns large private chambers testified to his many years on the bench. Floor-to-ceiling bookcases were lined with law books and journals. The dark oak paneling was virtually covered with numerous awards and photographs of famous and powerful people he had associated with during his tenure. His massive gold-trimmed desk, that had once belonged to a governor of California, was covered with even more awards and mementos from various bar associations and charitable organizations. Although he was only 55, his hair was now totally white, and his face was tanned and wrinkled from many years of outdoor activities. It was a stern face that easily commanded respect. He was reviewing a draft of an article he had written for a prestigious law journal when his secretary quietly knocked on his door and entered.

"Your Honor, Mr. Blackmon and two other gentlemen are here for your 11:00AM appointment."

Judge Burns looked at his watch. "Oh, yes, please send them in."

Mike Blackmon was an assistant prosecutor for Los Angeles County and a long-time friend of Phil Conley. They had met in law school and had kept in contact over the years. Blackmon had an enviable record and was known around the department as a "blue flamer" who would likely become some county's district attorney in the near future. With his dark hair and eyes, graying temples, and impeccable suits, Blackmon easily fit the mold of a successful courtroom litigator. As they walked over to Judge Burn's desk, he stood up and walked around to greet them.

Mike Blackmon was a few steps ahead of Ryan and Phil. "We really appreciate your seeing us on such short notice, Your Honor."

"Quite all right." They all shook hands.

"This is an old friend of mine, Phil Conley. He's an assistant district attorney in Houston. He is here trying to help Adam Daniels prove his son's innocence. And you know Ryan Hughes, of course."

"Pleasure to meet you, Your Honor."

"Please have a seat." Judge Burns returned to his chair as the visitors settled into large, comfortable chairs in front of the desk. Burns picked up a piece of paper and scanned it.

"Mr. Blackmon, as I understand it, Mr. Hughes and Mr. Conley are asking for permission to bring another video camera into the courtroom and some other equipment?"

"Yes, Your Honor. You have already agreed to allow several TV network cameras in the courtroom, and the State has no objection to another camera."

"What would be the purpose of another camera? Couldn't you obtain a copy of whatever you need from the news media?"

Phil answered for Blackmon. "We are attempting to input a new type of data from this camera into a personal laptop computer. A special software then analyzes the data in real time and provides a type of feedback that would help guide the defense attorney in subsequent questioning."

Judge Burns frowned. "What type of data?"

"It's rather complicated, but as we outlined in the request, it involves a new approach to measuring a person's physical reaction to questioning. One aspect involves voice-stress analysis."

Judge Burns suddenly sat up right in his chair, agitated. "A lie detector? Not in my court, Mr. Conley. As an assistant DA, you should know better."

An alarm bell rang in Phil. He had to diffuse the situation quickly. "This system is radically different, Your Honor. If I may, I would like to summarize the differences. The jury will not be aware of the data gathering or the information the system generates. The results will not be admitted as evidence, and the witnesses will not be confronted with the information generated during questioning. The information it produces will only be used to guide the defense attorney's cross-examination. It may prove of no use in a trial, but we are hopeful."

Burns glanced at Blackmon. "I find it strange the State has no objection."

"They gave me a short demonstration of the potential of the software, and it's a pretty radical approach to solving crimes. If it truly helps uncover some pertinent information in this case, then we are agreeable, Your Honor."

Phil leaned forward in his chair. "We have also agreed to share with the State all the information that is generated by the system, at the instant we have it, Your Honor. It will have no affect on the proceedings."

Judge Burns picked up a pencil and tapped on the desk as he reviewed the written request. "If this can be accomplished with no impact on the court or the proceedings, then I see no reason to not allow it. However, all of your equipment must be located in the back of the courtroom, out of sight of the participants."

Phil started to object at putting the equipment so far from the witnesses, but Judge Burns was already signing the request.

Judge Burns glanced up at Phil. "I would also like to know if the software has any affect on the outcome of the trial."

"We will make that known to the State and you in a special summary report at the end of the trial, Your Honor."

"Very well, is there anything else?"

Blackmon seemed pleased at the outcome. "No, Your Honor. Thank you for your time."

Phil, Ryan and Blackmon shook Judge Burns hand. As they left, he sat down to finish his journal article.

CHAPTER 21

Allen and Adam were waiting in interrogation room 5 to interview Ross again. Allen was performing a few system checks when Phil and Ryan entered. Allen immediately knew from Phil's expression that something was wrong.

"Well, did he agree to let us use the system, or not?"

"Yes... and no."

Adam's expression revealed his concern as Allen rubbed his eyes and sat back in his chair with a deep sigh. "What did he say?"

"We explained the system, and he was agreeable that it could be used..." Phil's voice trailed off and he sat down on the opposite side of the table from Allen.

Adam's voice was barely audible. "But... what?"

"All of the equipment has to be located in the back of the courtroom, out of sight of the courtroom participants."

Allen didn't say anything as he gazed off into the distance trying to evaluate the possible impact on the operation of the system.

Adam shook his head, fearing the worst. "Is it bad, Allen?"

Allen stared at the laptop screen. "It's not good. Not good at all...."

"How bad is it?" asked Phil, almost afraid of the answer.

"Well, let's go over it. The directional microphone could pick out a conversation in a crowd at 20 to 30 feet. What kind of distance are we talking about?"

Ryan picked up a pad of paper and started drawing a sketch of the courtroom. "The judge's bench and witness stand are here, and the TV cameras will be here... about a 100 feet or so."

"The courtroom will be relatively quiet," said Phil. "Won't that help?"

"Perhaps. But we could pick up some stray conversations among the spectators that could interfere."

"What are other potential problems?"

"The infrared camera has a limited range, but we might be able to find a telephoto lens to help some."

"Some?" asked Ryan.

"The change in facial blood flow is very subtle. I'm not sure what effect a zoom lens would have on the data already in the database."

The potential problems were beginning to dawn on Phil. "What other problems are there?"

"The biggest problem is the laser and spectrophotometer. At 100 feet, it will be extremely difficult to maintain an alignment of the laser to the witness's exhaled breath. And... I don't know if the spectrophotometer's resolution is great enough to pick up infrared florescence at 100 feet."

They were all silent for a moment as the magnitude of the problems became apparent.

"Theoretically," he continued, "at that distance, you could get interference from a woman's perfume in the audience."

Adam had to ask the most obvious question. "What if the laser part doesn't work? Can the rest do the job?"

"Probably not. All the data in the database is based on it. The advice data would be of marginal value."

Adam put his head in his hands for a moment. "Is there anything that can be done?" There was a hint of desperation in his voice when he looked up. "Phil, is there any chance of getting the judge to change his mind about the location of the equipment?"

Phil shook his head. "I don't think so. Judge Burns looks like someone who doesn't change his mind very often."

Adam stood up and started to pace the room. "Would better equipment help?"

Phil laughed. "We already have the best that money can buy."

"No, we don't." The others in the room turned to look at Allen.

"What?" asked Phil. He had written some pretty large checks to build the second prototype, and Ann had funded even more to miniaturize everything.

"Well, I'm probably not supposed to know about it, but the defense department has some pretty exotic imaging stuff in their satellites. It's definitely better. But we probably couldn't get our hands on it."

Adam was standing near Allen and put his hand on his shoulder. "I have a lot of friends that owe me favors. Tell me what you need."

Phil and Allen exchanged quick glances as Ryan chuckled. Allen picked up a notepad and started writing.

A little while later, Phil and Ryan were reviewing Ross› previous statements while Allen entered data into Sherlock. There was a knock on the door and Ross entered with a police officer who took up a guard position by the door. Ross shook their hands and sat down.

"How is the data gathering going, Allen?"

"Pretty well. We should have all the pretrial data in before the trial begins.

"So what is it you guys want?"

Phil picked up his notepad. "We just have a few more questions to ask you."

"Fire away."

Ross' cavalier attitude annoyed Allen and especially bothered Phil.

"Ross, I need to be blunt. You are on trial for murder, and we don't have a very strong defense at this time. The State has a lot of circumstantial evidence, and forensic experts to back up that evidence."

"And...."

Phil stared into his eyes. "You aren't being totally straight with us. We believe you're innocent, but the data indicates you're trying to hide something."

Ross suddenly seemed defensive. "I don't know what you're talking about."

"I think you do," replied Phil. "We don't know what it is exactly, but it seems to be drug related. Maybe you really did see Bill making a deal, or you do know something about the cash."

Ross didn't answer, increasing Phil's frustration. "Ross, you are on trial for your life. At best, you could spend most of the rest of your life in jail if you are found guilty."

Ryan put his hand on Ross' shoulder. "Ross, we are trying to help you."

Ross struggled to answer. "All right. The only thing I didn't tell you was that I took $10,000 out of a safety deposit box for Bill Kennedy the day of the party and gave it to him when I arrived."

Allen's jaw dropped and Phil sat down heavily, staring at him in disbelief. Only Ryan seemed to show no emotion. Ross hung his head.

"Why, Ross?" asked Phil in a barely audible tone.

"Bill gambled a lot and owed George's father quite a bit of money. He said George would be at the party, and he needed to pay him something until he could come up with the rest." Ross paused. He could tell from their expressions they didn't believe him. "I was just doing him a favor. I really only brought him ten thousand, not a hundred thousand. I don't know where that came from, or how it got in my car, and that's the truth."

Allen looked up from the monitor. "Yes, it is."

"Why didn't you tell this to me or your dad before," asked Ryan, his voice rising. "I'm trying to help you! How can I help you when you don't tell me the truth?"

"I'm sorry."

Allen and Phil exchanged glances. This was the first time they had seen Ryan show any real interest in the case.

"Why didn't you tell me this sooner? I would have put a detective on it." Ryan struggled to remain calm.

"Because... it didn't seem to have anything to do with George's murder, and I didn't want to get Bill in trouble with his dad over this."

Phil sat back in his chair angrily. "But this may give us a starting point on figuring out how it got into your car."

Ross hung his head again.

"Is there anything else you're not telling us?"

"Yes. I saw Bill and George talking together shortly after I gave Bill the money."

"What exactly did you see?"

"I didn't see much. I just heard them say the word 'thousand' several times and assumed Bill was trying to buy enough cocaine for everyone at the party."

Phil saw Allen frown. "What is it, Allen?"

"I don't know much about the street price of drugs, but wouldn't $100,000 buy a lot of cocaine? More than the people at that party would need."

Ross saw them waiting for an answer. "Are you asking me? I don't buy cocaine!"

"Never mind," said Allen. "What happened next?"

"George left and came back with a friend and they went upstairs. I just assumed they were making a deal there."

Phil scribbled in his notepad. "Was either one carrying a bag?"

"No, but it was cool that evening and both of them had jackets on. It could have been in one of the pockets."

Phil stood up, pacing the floor. "But you really don't know they were making a deal since you didn't actually see any drugs."

Ross seemed to mull that one for a while. "That's right. I didn't actually see them make a deal."

"If there had been a deal, what do you suppose happened to the cocaine?"

"Bill probably flushed it down the toilet when the police showed up. He used to do that with marijuana in high school."

"That's consistent with Rockey's and Dunlop's statements, and it explains why there were no drugs found in the house. Is there ANYTHING else you can think of that could help us prepare a better defense?"

"No. I still don't know exactly what happened that night. But I know I didn't kill George."

"Okay, that's all for right now. If you do think of something, please let us know."

"I'm going to have to tell Adam this." Ryan stood up to leave with Ross. When they left, Phil stood up and walked over to a window in the room and stared out, thinking. Allen corrected some words from the conversation just recorded and began an analysis of the data. The laptop beeped at him and "Allen?" appeared on the screen. Allen turned the speakers on.

"Yes, Sherlock."

"Allen, several statements of Ryan Hughes were less than truthful."

Allen and Phil exchanged shocked expressions.

"Ryan?" they both exclaimed together.

Allen and Phil spent a long-time discussing Sherlock's revelation about Ryan.

"He could be wrong," commented Phil. "After all, the analysis is based only on voice stress, and that's not always reliable."

"The factor was less than 20%."

Phil walked to a window and stared out at the lights of the city and the gathering darkness. He looked at his watch.

"It's almost 7:00PM. Adam will probably be home." He sat down at Ryan's desk and dialed Adam's private number. It was awkward at first, but he managed to lay out the evidence against Ryan in a reasonably coherent manner. There was a long pause on the other end as Adam tried to digest it all.

"I just can't believe Ryan is lying. Sherlock must be wrong."

"Allen and I discussed this, and we think Ryan is deliberately trying to sabotage this case."

"But why would he do that? This is probably the most famous case he'll ever work on?"

"I plan to ask him tomorrow."

Ryan's betrayal had incensed Adam. "I want to be there."

Michael Stone pressed the STOP button on a huge multi-unit DVR recording unit. He made an entry in a time log as the three men who had been intently listening to the dialog between Adam Daniels and Phil sat back in their chairs to reflect on the conversation. Already crammed with equipment, the presence of the three 'clients' made the interior of the van almost unbearable, even with the rear air conditioner running on full blast. They had been there for hours listening to Phil, Ryan, and Allen's pretrial defense effort. Stone jumped when Brian Limpanatti suddenly slammed his fist down on the equipment table and yelled.

"Dammit! We've got to get that system before it screws up everything!"

Michael waited until Brian had calmed down some and asked a question that had bothered him for some time.

"What does this murder trial stuff have to do the SEC investigation of Adam Daniels?"

Brian glared at him for a second. "That's classified," he growled. He pulled his cell phone out and quickly called Ryan. "They're on to you!" he exclaimed when Ryan answered. Michael couldn't make out the other side of the conversation, but he could see the tension building in the three men. Brian only replied. "Yes, I'll take care of it." He hung up and motioned his companions to leave. Michael was greatly relieved they were gone. He had seen the handle of a shoulder pistol under Brian's jacket. Were SEC agents authorized to carry guns? He would ask his boss the next time he called.

Phil, Allen, Adam and two uniformed officers payed an unexpected call on Ryan's office early the next morning. Ryan's secretary was flustered at the entourage but quickly unlocked Ryan's office door when shown a search warrant. She was as surprised as the rest to find the office almost empty. All the case records were missing, as were all of Ryan's personal items.

"I just can't believe it," cried Adam with anguish in his voice. Phil dismissed the officers. They all sat down to digest the unexpected turn of events in the last 24 hours.

"What do we do now, Allen ?"

"Luckily, we have all the key information stored in Sherlock's database."

"Phil, could this mess finally convince Judge Burns there's enough reason to delay the trial a few weeks?"

"I already thought of that. I called Mike Blackmon yesterday evening and he called Judge Burns to explain what had happened. Burns denied even this request for a delay, citing the fact that we've been on the case for a while and the two other original members of the team are still available. So, it looks like we have no choice but to continue."

Phil's choice was clear. "I'll be happy to take over the lead defense role if you don't mind, Adam."

"Of course not. In fact, I don't know what I would do if you couldn't. You are really a lifesaver, Phil... and you too, Allen. I won't forget this." Adam was almost in tears.

"We have a lot to do, Allen. Jury selection starts the day after tomorrow."

Allen cracked his knuckles. "I'm ready."

The murder trial of Ross Daniels opened with a great deal of fanfare and media attention. Adam Daniels made every effort to avoid the press and was happy Phil was used to the constant crush of the media. Phil diverted as much attention away from Adam and Ross as he could.

Phil was surprised at how knowledgeable and helpful Ryan's assistant, Jim Robbins, and his paralegal, Lucy Stubbs, were. They had been extremely frustrated at Ryan's lack of effort to form a credible defense. Even the two detectives Adam had hired to help in the investigation confided they felt Ryan had been a bigger hindrance than help in Ross' defense. Phil quickly took control and put them to work. Adam was more than eager to provide Phil with any additional resources he requested.

A jury of six men and six women was selected in record time. Adam lamented the fact that Allen's lie-detector system was not available during the selection of the jury, but Phil was pleased with the makeup of the jury and Allen assured him the system would be ready when the defense portion of the trial began.

Mike Blackmon opened the State's case with a strong and convincing argument for Ross' guilt. Over the next several days, the prosecutors paraded a string of witnesses and experts through the court to substantiate its case. With the limited amount of defense data available, Phil could only try to raise reasonable doubt in the minds of the jury as each fact was presented. Adam's anxiety increased with each passing day as the prosecution's case was presented.

CHAPTER 22

Allen sat back in his chair and rubbed his eyes. Five straight hours of entering data in Sherlock's database had exhausted him. He hadn›t even noticed that the early Saturday afternoon had faded into night or that he hadn›t eaten since breakfast. He hated weekends. At least he had a fairly routine schedule during the week. As he entered the last of the data and started Sherlock's data analysis mode, he saw a window open on the screen with a notation from Sherlock that he had dialed into a crime database and was cross-checking some information. He glanced at his watch and suddenly remembered he had promised Alicia he would go to a fancy investor's party in less than an hour. He stood up and ran to the bathroom. As he was dressing, the laptop began beeping and "ALLEN?" appeared on the screen.

What now? He reluctantly turned on the laptop speakers.

"Yes, Sherlock?"

"Allen, a cross check of the State of California's crime database finds a match with another individual in my database. Do you want to review the information?"

I'm going to be late. He started to say no but reconsidered. Maybe he should check it out. "Yes, proceed."

A window popped open with an arrest record. "The criminal file of George Artoles shows an arrest on June 23rd of last year for fighting in a bar."

Allen stared at it but didn't see a connection. "So, what?"

"A Gene Abrams was also arrested at that time."

Allen still didn't make the connection. When he didn't reply, Sherlock continued. "It may be a coincidence, but a search of the national crime database indicates that Gene Abrams had an extensive prior arrest record in Houston." Sherlock waited, then continued. "One of the individuals interviewed for the Stevens Equipment Company has the same name and city address... the street address is different, however. Here is the arrest photograph."

A picture of Gene Abrams appeared on the screen with almost a crew cut and a very short, cleanly manicured beard. He didn't recognize Gene at first but stared in disbelief at his resemblance to Ross and the address listed... it was his address when he was still married to Mary! A connection suddenly dawned on Allen that went beyond the surly Gene Abrams he had seen in the interview room at Ann's company. He had seen this man at his final divorce proceedings. He sat down stunned. Pieces began to fit together.

"Excellent work, Sherlock. Thank you."

Allen could almost detect a faint arrogance in Sherlock's reply. "Elementary."

He grabbed the receiver of the phone and quickly punched in Phil's number.

It seemed to Allen he had barely hung up the phone when Phil knocked on his door. He laughed as he opened the door. «I didn›t know you could fly.»

Phil ignored the comment as he rushed in past him to the laptop. "Let me see the picture, Sherlock!"

This was the first break Phil had in the case and his energy was at a low point. He needed this. When Gene's picture appeared on the screen, he involuntarily gasped in surprise.

"Jesus Christ! He looks like Ross."

Allen stood next to him studying the photograph. "With the right clothes...."

Phil glanced at Allen then stared at the screen "If we could just see the current Gene Abrams like this...."

Allen stared at the screen for a second and then grabbed the hotel phone and dialed a number. He pressed the speakerphone button when she answered.

"Ann, did you use a digital camera to take the application photos for the employment interviews? Can we get a copy of one of the pictures? Could they email it to you?"

"Slow down, Allen. The answers to your questions are yes, yes, and yes. What's going on?"

Allen quickly explained their discovery and the link to Gene.

"I'll call Amanda and ask her to find Gene's photo."

"Can you find out the picture format as well? I think most image files can be inserted into face-making software."

"I'll find that out too. We used a digital camera, so I'm sure she can attach it to an email and send it to me. I'll call you be back shortly."

Allen hung up the phone.

"What are you going to do?"

"I hope to show that Gene Abrams could have been the killer. Do you think you can arrange a session with a police sketch artist familiar with the face-making software?"

"I'll get Jim Robbins to do that right now." After a brief phone conversation with Robbins, he stood staring at Gene's picture for a moment.

"Allen, during the interviews at Ann's, you said you thought you had met Gene Abrams before. Did you ever remember where?"

"Yes, he was in the back of the courtroom a few times at the end of my divorce trial. I remember wondering why visitors watched divorce trials that didn't concern them. But I sort of forgot about it with all the other things going on. Now we just need the interview picture of Gene Abrams."

"I'm going to go get my trial notes. I'll be right back," Phil said as he closed the door behind himself.

Allen looked at his watch and remembered his date with Alicia. He laughed. Maybe Ann would like to go in his place. Surely, she would appreciate the opportunity to meet some powerful investors. He picked up the phone to call her.

Allen was typing on the laptop as Phil entered.

"Allen, Ann gave me this disk for you before she left for the party. She said she just got the picture in an email from Margaret. By the way, Margaret says, 'Hi.'"

"She's one fine lady."

"You'll never guess what Ann told me... Margaret is dating Joe Willis!"

Allen's jaw dropped. "The enormous Joe Willis with the hot temper?" They laughed loudly.

"That's the one."

"What an odd combination! Maybe she can straighten him out."

Allen inserted the disk into the laptop and started typing. A moment later Gene Abrams' employment application picture appeared on the screen. "This may be just what we need."

"While you're doing that, I'm going over to Adam's house and tell him what we've found."

Phil walked out the door just as Alicia walked in. Allen's surprise showed.

"I thought you and Ann were going to that big investor party?"

"She's waiting in the limo. We want to be fashionably late. Ann told me you may have found something?"

"Yes. Come here. I want to show you something."

"I've already seen it."

Even though they were alone, Allen couldn't help blushing. "On the laptop!"

"Oh."

She stood next to Allen, putting a hand on his shoulder. "Who is that?"

"He might be the killer."

Alicia looked at Allen and then at the image again. "He kind of looks like Ross."

"Great! That's what I wanted you to say. I hope to get the witnesses to say that."

Allen stood up and gave her a quick hug.

"Don't do that... unless you mean it," she said teasingly.

He started to kiss her, but she pushed him and he fell back into his chair. She sat on his lap, straddling him. "Now, what did you want to show me?"

"Isn't Ann waiting for you?"

Alicia looked at her watch. "I still have ten minutes."

"I only need five."

Alicia chuckled as Allen picked her up. "Just don't mess up my hair."

Allen guided the police sketch artist through a series of changes to Gene Abrams› picture. Phil watched quietly from a nearby chair, fascinated at the changes appearing on the laptop screen as the artist maneuvered the mouse around on the table.

"Can you make the hair shorter? Almost a buzz?"

"Sure."

Allen suddenly stopped to stare at an image that looked very much like Ross.

"Can this software morph one image into another one?"

"Watch this!"

The interview picture of Gene slowly transformed into the clean-cut image and then into Ross' arrest picture.

"That's perfect. Great job!"

"Thanks."

"Phil, look at this."

Phil moved closer to watch the morphing of the pictures.

"What do you think?"

"Wow! I'll talk to Mike about showing this to the witnesses. Great job, guys!"

He left quickly. The sketch artist saved the files and picked up a form to file an evidence report.

Allen muttered to himself, "But will it convince a jury?"

CHAPTER 23

The door to Allen's room slowly opened as Brian and Mario entered as quietly as possible. They quickly looked through the closet, the dresser, and the desk. A few moments later, Juan walked in and stood at the door looking up and down the hallway.

Brian suddenly whispered loudly, "Here it is!" as he pulled the small laptop from a dresser drawer. Mario quickly stood next to him.

"Let's get the hell out of here!"

Juan quickly walked inside and closed the room's door. "He's coming!"

Brian whirled around in surprise. "You idiot! You were supposed to wait in the lobby and let us know when he came in. How are we going to get out of here now?"

Brian quickly turned off the lights and stood behind the door with a gun raised. Mario slipped into the bathroom, and Juan walked quickly to the bedroom with a gun in his hand. There was the sound of a card key in the door. Allen walked into the room, and immediately Brian pressed his gun into Allen's back.

"Don't move, and you won't get hurt."

Allen raised his hands.

"Take your wallet out and drop it on the floor."

Allen did it without hesitation. "What do you want?"

"Shut up and put your hands on the wall."

Allen did as he was instructed and Jack grabbed a wooden chair from the desk and pushed it under him.

"Now sit down and put your hands behind your back."

Allen sat down and put his hands behind the back of the chair. Juan and Mario came out and stood behind Brian.

"Now tie his hands behind his back."

"With what?" replied Mario. "I didn't bring any rope."

"Tear the sheets up and use them."

Mario and Juan started tearing the sheets into strips. They tied Allen's ankles to the legs of the chair and his arms behind the back of the chair, and then tied some strips over his mouth.

"Okay. Let's go," hissed Brian.

They left quickly, not bothering to close the door. Allen saw Juan carrying the laptop. Almost immediately, he leaned forward and swung the chair around smashing it against a chest of drawers. He stood up and pulled the broken chair legs from his ankle wrappings and kicked the wrappings off. He used a chair leg to break the mirror above the dresser and used a piece of glass to cut the sheets binding his hands as he started to run toward the stairs. Less than 20 seconds had passed since the laptop had been taken.

Juan was examining the laptop as Brian, Mario and he were riding the elevator down. The elevator stopped at the tenth floor. As the doors opened, a male guest started to get on. Brian pushed him back off the elevator.

"Take the next one!" He pushed the close button and the elevator started back down. It stopped again on the eighth floor, and the doors opened. Two elderly ladies started to get on, but Brian put his hand up to block them.

"Excuse me ladies, this elevator is going for repairs. Please take the next one. Thank you."

He pushed the close button and the elevator started down again.

"This sucks," said Mario. "We should have gone down the stairs."

"Shut up," replied Brian. "We'll be down in a minute."

Allen almost flew down the stairs, touching only one in five or six stairs as he descended. He burst out of the stairwell in time to see the elevator stopped at the eighth floor. He saw the housekeeping closet nearby and ran inside. He found a mop and broke the end off. He put the handle behind his back and walked quickly to stand next to the elevator that had just arrived.

The door opened, and Brian walked out first. Allen tripped him with one end of the mop handle and hit him on the head with the other end in almost a single motion. Mario was right behind Brian and Phil hit him in the stomach. He tripped over Brian and Allen hit him on the head. Juan was holding the laptop and was trying to get a gun out of his pants and almost had it as Allen entered the elevator. He kicked the gun out of his hand, and put the mop handle under his nose, pushing him back against the wall of the elevator.

"Hand that to me, or I'll give you a nose job."

Juan was so scared he raised his hands, dropping the laptop. Allen leaped and caught it, body-slamming Juan against the back wall of the elevator at the same time, knocking him out cold. Allen twisted in mid-air landing on his back. As he got up, he picked up the mop handle and walked out to stand over the other two. He noticed a couple standing a few feet away, staring in disbelief.

"Please get the hotel's security."

As the man rushed off, Allen turned the laptop on. It gave the normal startup beeps and Allen breathed a deep sigh of relief. A security guard ran toward him with the male guest as Allen dropped the mop handle.

Phil, Allen, and Ann had been invited to a meeting with Detective Johnson and were delighted at his news.

"The three individuals you stopped last night have refused to talk, but we did a little investigating and traced them back to Jake LoBlanco. He's a powerful underworld boss in Las Vegas."

Ann frowned at the mob connection. "What does he have to do with this?"

"His daughter, Maria, was married to George Artoles."

The link to LoBlanco caught them by surprise.

"So he's out for revenge, as well as Antonio Artoles?"

"It would seem so, Ann. Antonio's operations seem to be pretty much limited to LA. Apparently, Maria met George several years ago when there was some friction between Jake and Antonio. They probably had some overlapping areas in dispute and George met Maria when he acted as an intermediary. George and Jake seemed to have resolved their differences when their children married."

"What happens now?"

"We issued an arrest warrant for Jake this morning in connection with the robbery attempt last night. But he has a lot of powerful friends in Nevada, and it won't be easy getting him out. Even so, Adam Daniels knows some pretty powerful people as well... so we might. We checked with some of the hotel staff, and those three clods have been practically living at the hotel since you all arrived."

Phil rolled the possibilities around in his head. "How do you think they found out about Sherlock?"

"It's still just speculation at this point, but we think several of the partners of the law firm Allen Daniels contacted to defend his son were indebted to Antonio Artoles, and he may have used his influence to get Ryan Hughes appointed as Ross' attorney. Ryan must have told them."

The connection to Ryan seemed logical to Ann. "So that's the connection! You were right about Ryan, Phil."

Dave gave Phil a questioning look.

"Oh... I had commented the first day I met him that he seemed a little inexperienced for such a case with such notoriety."

Ann corrected him. "You used the term 'in over his head'."

"He just had a different agenda than we thought."

Allen looked puzzled. "But we didn't tell Ryan about Sherlock, only the lie detector system."

"He must have found out somewhere else."

Dave scratched his chin. "Antonio undoubtedly has informers here in the department. It's amazing though, because only a few people know about your software and what you're trying to do."

"Wouldn't that make it easier to find out the informer?"

"Possibly, Ann. We have asked Internal Affairs to begin an investigation. By the way, Mr. Atkins, we showed the witnesses your animation of Gene Abrams transforming into Ross Daniels. Several of them agreed the person they claimed to have seen running down the hall could have been Gene Abrams, so now we have an interest in talking to Mr. Abrams as well."

"I spoke with Mary last night. I told her we had finally put the pieces together and we knew it was Gene who beat her up. It took some convincing, but she finally agreed to press charges against him."

"Mary gave the police his address and they arrested him this morning," commented Ann. "He probably would have gotten out on bail, but Phil explained to the detective in charge that Gene was wanted out here for questioning on a murder case. Adam had an extradition request ready for a judge's signature this morning, so Gene is probably on a plane out here right now."

Dave stood up and poured himself a cup of coffee. "You are a remarkably efficient team."

Allen poured himself a cup of coffee. "Let's just say we have an interest in seeing if Mr. Abrams is the person responsible for the murder."

"When do you plan on putting him on the stand?"

Phil was looking at a calendar on the wall of Dave's office. "The defense phase of the trial may begin as early as tomorrow. I think Mr. Abrams may turn out to be a star witness for us."

"That's excellent. By the way, has your software system been of any help so far?"

"It flushed out Ryan Hughes's deceptions, and of course, led us to Gene Abrams. Also... based on past cases in the database, it advised that there may be a physical resemblance between the person identified by witnesses and the actual killer."

They all laughed.

Dave shook his head in amazement. "Anything else?"

Allen was making some notes to himself as he answered. "This morning, Sherlock asked why the police arrested an 'unconscious' suspect."

Phil gave him a questioning look. "Unconscious?"

"I think the prevailing impression that Ross was found sleeping came from one of the witnesses the reporters interviewed that night. I cross-checked the police report, and the arresting officer's report described Ross as 'unconscious' rather than 'sleeping'."

"I hadn't noticed that."

"Neither had we."

CHAPTER 24

The defense portion of the trial began unceremoniously compared to the prosecution's exposition before the network cameras.

They had not received the super-sophisticated equipment Adam had somehow miraculously obtained until the prosecution's case was almost concluded. When the equipment finally arrived, Allen had chosen to spend the time until the defense portion began by testing the new components.

However, Sherlock's new spectrophotometer and infrared video camera were now situated in the rear of the courtroom as required by Judge Burns' orders. Both were pointing at the witness stand. A new highly sophisticated tracking device had been obtained 'on loan' from a defense contractor. It easily kept both devices aligned on the witnesses' breath. The image resolution of the new infrared video camera, also borrowed from a defense contractor, amazed everyone, even Allen. Both items had been quickly integrated into Sherlock's data-gathering system and tested on members of the defense team and some of Adam's domestic staff. Both appeared to pass the tests with flying colors.

Allen had added a wireless transmitter to the laptop's speaker output so Phil could hear the results of Sherlock's analysis through an earphone that looked remarkably like a hearing aid. They had provided a similar receiver to Mark Roberts, an assistant attorney on Mike Blackmon's prosecution team, in lieu of a display monitor as originally planned.

Gene Abrams obtained a lawyer and had managed to block his extradition to California. Adam flew to Houston with a new team of lawyers that specialized in extradition and was confident they would have him on a plane in a few days. Phil began the defense by recalling some of the prosecution's witnesses. He hoped the system could point them in a new direction and uncover the truth around George's death.

Judge Burns› steel gray eyes were focused on Phil. «Call your first witness, Mr. Conley.»

"The defense would like to recall Mr. David Jones to the stand." The packed courtroom buzzed a little and Judge Burns pounded his gavel for silence as Jones made his way to the stand. Jones was one of the five witnesses that had identified Ross leaving the murder scene. Based on an analysis by Sherlock of his videotaped interview with the local TV station, there was an indication he might not be telling the truth, and Phil wanted to know why. Judge Burns reminded Jones he was still under oath.

"Mr. Jones, you previously said you saw the defendant leaving the murder scene and running down the hallway. You also later picked him out of a police lineup. Are you certain the defendant, Ross Daniels, is the person you saw?"

Jones appeared completely at ease. "Yes, sir."

To Phil's surprise, Sherlock commented, "The probability of truth is 15% for the last response of witness Jones." Phil glanced at Mark Roberts who seemed just as surprised. Phil stole a glance in Allen's direction. He was scratching his head, confirming the analysis, indicating a definite lie. But which part was a lie? "Mr. Jones, do you know the penalty for lying under oath in the State of California?"

Jones face revealed a sudden terror. He stumbled for an answer. "Err... yes, I guess so."

Several possibilities raced through Phil's mind. Jones could know who the actual murderer was, or he could know Ross wasn't the murderer. He could even be lying about seeing anything at all. Phil decided to try the last possibility first.

"Let me rephrase the question a little. Would it be accurate to say that you saw a man leave the scene and run down the hallway?"

Jones appeared to be confused at the difference, but answered, "Yes, sir."

Sherlock identified that response as a lie. To Phil, it appeared Jones was simply lying and probably didn't see anything. But why lie and say he did? He decided to throw Jones off a little more then hit him again.

"Mr. Jones, your name doesn't appear on the party invitation list. Were you invited to the party?"

Another wave of terror was reflected in Jones' face.

"Uh... no. Well, not exactly."

"Not exactly? Were you invited to the party or not?"

"No. One of the guests asked me to go with him."

"Oh... which guest, Mr. Jones?"

"George Artoles."

Most of the spectators didn't react to Jones's statement, but it caused quite a stir among the prosecution team. They were clearly unhappy that a key government witness was brought to the party by the victim. What was the connection?

"How did you know Mr. Artoles?"

He turned to Judge Burns. "I'm not on trial here, Your Honor. Do I have to answer all these questions?"

Judge Burns glanced at the prosecution team before replying. "I don't seem to hear an objection from the prosecution. The witness will answer the question."

Jones replied reluctantly. "He was my bookie."

That comment caused a stir and Judge Burns pounded his gavel for silence.

"Mr. Jones, did George Artoles bring you to the party to lie and say you saw something when, in fact, you didn't?"

Jones face went white. "I... I don't know what you are talking about."

Phil saw Allen scratching his head again. He didn't need Sherlock's confirmation to know Jones was lying. Why would George bring someone to the party to lie about something that was going to happen?

"One last question... and remember, you are under oath. Had you ever seen the defendant before the trial began?"

He could see Jones fighting an internal conflict on whether to tell the truth or not. The truth finally won out and Jones answered in a whisper. "No, sir."

"That's all we have for this witness, Your Honor."

Judge Burns stared at the prosecution table. "Gentlemen?"

Mike Blackmon wearily stood up. "The State requests Mr. Jones be held on possible perjury charges, Your Honor."

"Thank you, Mr. Blackmon." He turned to the bailiff. "Mr. Jones will be held for 24 hours pending a charge of perjury in this case."

As an officer led Jones out, Phil called his next witness. "The defense would like to recall Bill Watson."

Bill Watson was the other eyewitness Sherlock had identified as not telling the truth. Having witnessed the results of David Jones' testimony, Bill Watson did not appear eager to testify. When he was reminded that he was still under oath, he barely muttered a response. Phil went on the offensive.

"Mr. Watson, like Mr. Jones, did you accompany another guest to the party?"

Watson was sweating profusely as he replied. "I refuse to answer on the grounds that my answer may tend to incriminate me."

The message from Sherlock was a less direct. "Voice Stress Analysis and facial blood-flow pattern indicate internal stress."

Phil didn't need Sherlock to see the stress Watson was under. Sweat streamed down his face. "Mr. Watson, did you accompany Mr. Artoles to the party?"

Watson repeated his desire to claim the Fifth Amendment.

"Mr. Watson, do you still stand by your earlier statements, that the defendant, Ross Daniels, is the person you saw leaving the scene of the murder and running down the hall?"

Watson barely squeaked his claim to the Fifth Amendment.

Allen picked up a microphone and Phil heard his comment through his earphone. "I think he knows who the murderer is."

Phil tried repeatedly, but Watson refused to answer any additional questions. He finally gave up. Watson hung his head and shuffled quickly to the back of the courtroom.

Phil wiped his forehead with a handkerchief before continuing. "The defense would like to recall Mark Dunlop."

Mark Dunlop had made an agreement with the State to tell everything he had seen at the murder scene. In exchange, the State agreed not to prosecute him for his role in the drug deal that seemed to be at the heart of the matter. There had been no physical evidence at the scene that could be used against him anyway. Dunlop was reminded he was under oath as Phil approached him.

"Mr. Dunlop, I would like to ask you a series of questions about the murder, since only you and Sam Rockey were present...."

Dunlop interrupted him. "Bill Kennedy and the defendant were there as well."

Phil glanced at Allen, but he made no sign. Dunlop had not known Ross before the party, and he appeared to be telling the truth. At least he believed Ross was there. Hopefully, he was merely mistaken.

"Of course. I stand corrected." Phil paused. "Mr. Dunlop, did George Artoles bring you to Bill Kennedy's party to sell a large quantity of drugs?"

Dunlop was furious. "The State and I made a deal. I don't have to answer that."

"No... the State merely said they would not prosecute you for it. You still have to answer the questions presented to you here or you could be held in contempt."

Dunlop looked at Mike Blackmon for help but received none.

Judge Burns leaned toward the witness stand. "You will answer that question, Mr. Dunlop."

Dunlop fought hard to control himself. "All right. The answer is yes."

"Would you say that you and George Artoles were good friends?"

Dunlop seemed surprised. "No, I wouldn't say that."

"Just friends?"

"No."

"Casual acquaintances?"

"No. We were just involved in a few deals before this."

"As business partners?"

"No. Not exactly."

"Then why would George provide you with an opportunity to make quite a bit of money for something he could have done himself?"

Dunlop thought that one over for a moment. "I don't know, really. Maybe his regular supplier couldn't come through in time. I don't know."

"It almost sounds as if you and George were rivals for the same business."

"I wouldn't exactly say that, but it happened sometimes."

Phil pulled a TV on a rack in front of Dunlop. Several large-screen TVs had been positioned so the jury and the audience could see Phil's presentation as well.

"Mr. Dunlop, I'd like for you to watch a small demonstration of a face-making software package often used by the police to help eyewitnesses generate a likeness of an individual wanted for questioning or a crime."

Dunlop didn't say anything but watched the heavily bearded interview picture of Gene morph into the clean cut arrest photo taken around the same time as the murder. The arrest photo of Gene then morphed into Ross' arrest photo. The morph sequence was repeated a few times and Dunlop sat back in his chair and absently pulled on his chin, thinking.

"Mr. Dunlop, you said you had never seen Ross Daniels before the party. Is it possible the individual you saw in that demonstration could have been the person you heard George Artoles call 'Ross'?"

"I guess it's possible, but I'm not sure."

The courtroom erupted, and Judge Burns gaveled for silence. Mike Blackmon's team went into an immediate huddle.

"Now, Mr. Dunlop, I'd like to turn back to the murder for a moment. Earlier you said you and Sam Rockey tried to interrupt the fight between George Artoles and the man you assumed to be Ross Daniels. Is that not correct?"

"Yes."

"You also said the gun went off at this time, with the bullet hitting George in the chest."

"Yes."

"Why would you try to stop a fight between someone you considered a rival and someone you didn't even know?"

The blood drained from Dunlop's face but he tried to remain calm. "What?"

"Why would you risk getting harmed to break up a fight between two rather large men, neither of which was your friend?"

Beads of sweat appeared on Dunlop's face as he replied. "Err... I don't know. Maybe it was just instinct."

Phil almost jumped when Sherlock advised, "The probability of truth was 5% for the last response of witness Dunlop."

When Phil looked at Allen, he was smiling and scratching his head. Clearly, another lie. But which part? Phil glanced at Mark Roberts who was whispering something in Mike Blackmon's ear.

Phil spent the next hour trying to pin down the reason why Dunlop had tried to interrupt the fight but drew a blank. Judge Burns interrupted.

"Mr. Conley, it's now almost 4:30. Perhaps we can resume this line of questioning tomorrow?"

"Yes, Your Honor."

Judge Burns pounded his gavel. "This trial is recessed until tomorrow morning at 9:30AM."

The bailiff stood up. "All rise."

Everyone stood up as the Judge left through a side door. Dunlop almost ran from the courtroom with two officers in close pursuit. Phil stared off after him, wondering what he was trying to hide, until Allen touched his arm.

"Adam called a little while ago. They will have Gene on the first plane out tomorrow morning. We should be able to bring him here in the afternoon."

Phil looked relieved. "Thank God. Maybe we will finally be able to sort all this out."

CHAPTER 25

Judge Burns pounded his gavel to signal the start of the day's proceedings. He was a little surprised when Mike Blackmon stood up to address him.

"Your Honor?"

"Yes, Mr. Blackmon?"

"I must inform the court that Mr. Dunlop seems to have vanished. We checked every location he was known to frequent, but he seems to have slipped out of sight."

Judge Burns was not happy. "Wasn't he in protective custody, Mr. Blackmon?"

"Yes, Your Honor, but he eluded the officers assigned to him."

Judge Burns looked at Phil. "Mr. Conley, can you continue with other witnesses until Mr. Blackmon can find Mr. Dunlop?"

Blackmon winced at the sarcasm in Judge Burns' voice.

"Yes, Your Honor."

"Very well, then. Let's proceed."

Allen had just informed Phil that U.S. Marshalls were in route to the courtroom with Gene. He wondered if he could pry anything useful from Sam Rockey, the other witness to the shooting. "The defense would like to recall Sam Rockey to the stand."

Rockey was reminded by Judge Burns that he was still under oath.

"Mr. Rockey, did you see the face-generation software demonstration yesterday?"

"Yes, sir."

"Do you think it's possible the person shown in that demonstration could have been the person George Artoles called 'Ross' prior to their fight?"

"I'm not sure. Could I see it again?"

"Maybe still photographs would help." Phil held up the photographs for Judge Burns. "Your Honor, these photographs have been previously submitted as defense exhibits 24 and 25."

Judge Burns nodded and Phil handed them to Rockey.

"The first picture is of another suspect in the case, Gene Abrams. This picture was taken around the time of the murder. The second picture is of Ross Daniels taken when he was arrested. These are the pictures shown being transformed in the demo. Would you say it is possible that Gene Abrams was the person you saw fighting with George Artoles before he was killed?"

Rockey studied the photographs carefully. When he looked up, his face was expressionless. "It's possible."

Phil glanced at Allen, but he simply continued to stare at the data on the screens. Mike Blackmon's team held another quick huddle. The courtroom proceedings were suddenly interrupted as the rear doors burst open and two U.S. Marshalls dragged a reluctant Gene Abrams in and sat him roughly onto the first-row visitor's bench. The crowd buzzed in excitement when they saw his handcuffs and leg irons.

Gene was dressed in inmate clothing, but his hair and beard were even longer than they were at the interview in Ann's company. Gene must not have cut his hair since the murder. He had changed so much since the interview Phil almost didn't recognize him. Clever disguise. It would have been hard to find him without Mary's help. Several spectators and Phil laughed out loud at Gene's appearance until Judge Burns gaveled for silence.

Phil approached the judge's bench. "We have no more questions for this witness, Your Honor."

"You may step down, Mr. Rockey, but don't attempt to flee these proceedings as Mr. Dunlop has."

"Yes, Your Honor," Rockey said quietly and hurried to the back of the courtroom. Judge Burns signaled to two officers, and they took up a guarding position behind Rockey.

Phil walked back to the defense table and picked up his trial notes. "The defense would like to call Gene Abrams to the stand. And Your Honor, we would like the record to show that Mr. Abrams is a hostile witness."

"It will be so noted."

A Marshall standing next to Gene helped him up. Gene started to resist and both Marshalls gleefully helped pull and drag the reluctant witness to the stand. He appeared to be in a defiant mood. He was sworn in and sat down. He immediately recognized Phil as he approached, and scowled.

"Mr. Abrams, did you know the victim in this case, George Artoles?"

"NO!!"

Sherlock advised, "The probability of truth was 10% for the last response of witness Abrams."

"Did you attend the party hosted by Bill Kennedy approximately nine months ago, at which the body of George Artoles was found?"

"No."

Phil didn't need to know that the probability Gene was telling the truth was only 5%. "Let me ask the question differently. Were you in the Kennedy mansion the night of the party held about nine months ago?"

"No."

"The last response of witness Abrams has a truth probability of 3%."

"Let me remind you that you are under oath, Mr. Abrams."

"I know that," replied Abrams defiantly.

"Do you know the penalty for perjury in the State of California?"

Gene seems a little unsure of how he should answer. "Uh... yes, I guess so."

"Excellent! So let me ask the question one more time. Were you at the Kennedy mansion the night of the party previously mentioned?"

"I may have been, but I don't remember. That was a long time ago."

"What if we had witnesses that would be willing to say they saw someone there that night that matches your description, and when they were shown your picture, also agreed that you could be the same person seen running from the bedroom where the victim was found?"

"Then they're full of shit. They were probably drunk that night anyway!"

"I didn't say they had been drinking. Why would you assume they were?"

"Why, uh... you said they were at a party, so I assumed they were probably drinking that night."

Gene seemed relieved that he had fabricated a good excuse.

"What if we could produce a copy of a rental-car agreement signed on that date by someone that matches your description, and we also had a witness that identified you as the person renting the car?"

The detectives supporting the defense case had been busy.

Gene seemed a little uncomfortable at those questions.

"I didn't say I wasn't in LA that night."

"The agent at the rental counter also said you had a short haircut, were wearing expensive clothes, and rented a really hot-looking foreign auto."

A dark cloud rolled across Gene's face. "What about it? It's a free country, isn't it?"

"Yes, but is it possible you were dressed that way so you could get into the fashionable party being held at the Kennedy mansion without being hassled or even noticed."

Gene scowled at him again. "Anything is possible!"

"Mr. Abrams, we checked every bank in Houston and the surrounding area and could not find a checking or savings account in your name."

Gene seemed confused. "So what?"

Phil held up a piece of paper in front of Gene. "So how could you have managed to buy or rent so many expensive things? Your recent job application at Stevens Equipment Co. indicated you had not held a job in almost a year."

"I was here to help a friend out, and he gave me some money."

Phil put his hands on the railing in front of Gene, in an 'in your face' approach. "What was the nature of the help you were providing."

"That's none of your damn business."

"Your Honor, we believe the purpose of Mr. Abrams' visit to the Kennedy estate is key to determining what really happened that night. Would you direct the witness to answer the question?"

Burns glanced at Gene as he scribbled in his trial journal. "The witness will answer the question."

"No."

Burns was furious. "Mr. Abrams, you will answer the question, or I will hold you in contempt!"

"I refuse to answer that question on the grounds that it may tend to increment me."

Several onlookers laughed and Phil smiled. Gene was red-faced as he realized he had said something wrong.

"You mean incriminate you?"

"Whatever. I refuse to answer it anyway."

Judge Burns leaned toward him. "Mr. Abrams, I could hold you in custody UNTIL you decide to answer that question."

"I still won't answer it."

Burns started to challenge him to answer the question again, but was cut off by Phil.

"Mr. Abrams, did the assistance you were providing to your friend have anything to do with an illegal drug deal at the Kennedy estate."

"No!!!"

There was no response from Sherlock and Phil glanced at Allen, but he was resting his chin on his hands staring at the laptop monitor.

"Did it have anything to do with the host, Bill Kennedy?"

"Why don't you ask him?"

"We will... soon. But for now... did your business there have anything to do with Bill Kennedy?"

"No!!!"

There was no response from Sherlock and Phil was surprised Gene was not involved in the drug deal. Why was he there, then?

"Were you there to buy drugs from Mr. Artoles?"

Although Phil was only a few steps away, Gene shouted his answer. "I told you, NO! Are you deaf?"

"Not yet." Phil glanced at Allen, who was slowly shaking his head. They were on the wrong track.

"Were you there to sell him some drugs?"

"No!"

Allen was still slowly shaking his head. They still were on the wrong track.

Why was he there? Something else... maybe it was money related. "Were you there to pay off a debt?"

Gene seemed surprised at this question. He paused before answering. "Uh, no."

Sherlock confirmed Gene was telling a borderline lie. The answer had some element of truth and some falseness. Phil was desperately searching his brain for a new path to follow. He began to mutter out loud to himself. "Debts... credit-card debts... loan sharking... gambling debts...." He was staring off into the distance, when he suddenly stared into Gene's eyes. "Were you there to collect on a gambling debt from Bill Kennedy?"

"Uh, no." Gene's voice began to tremble and he developed a nervous twitch on one of his eyelids.

Another borderline lie confirmed by Sherlock. They were getting closer. Phil stepped closer to Gene. "Were you there to pay off a gambling debt?"

Gene looked desperately as if he wanted to leave the stand. "No! No! I wasn't."

Allen was scratching his head. A clear lie. But why would paying off a gambling debt lead to a fight with George? And Gene had admitted he didn't have any money. How could he pay off a gambling debt? Phil moved even closer and Gene backed up against the back of the witness stand.

"You went there to pay off a gambling debt by doing something for someone at the party. What did he ask you to do, Mr. Abrams?"

"I don't know what in the hell you are talking about!"

Sherlock advised of the lie, and Phil put his hands on the front of the witness stand again. "Two witnesses have said it is possible that you could be the person they saw get into a fight with George, and struggle with him before he pulled a gun that went off, killing him. Who were you doing a favor for, Mr. Abrams? Mark Dunlop?"

"No!!"

Strangely, there was no response from Sherlock.

"Did Mark Dunlop pay you to kill George Artoles to eliminate his competition?"

"No!!"

Something flashed in Phil's mind.

"Did Mark Dunlop see your resemblance to Ross Daniels and then instruct you to stage the fight so he could kill George Artoles?"

"No!!!"

There was no response from Sherlock and Phil almost lost his train of thought.

"Dunlop told us he struggled to take the gun away from the person resembling you and George. Did you force the shot to try and kill Dunlop?"

Gene was confused. The courtroom seemed to be spinning and the two U.S. Marshalls seemed to be moving in on him. Phil was practically yelling in his face.

"No, it was George's idea," he blurted out before he could catch his tongue. The audience gasped. Even the prosecution team seemed flabbergasted.

Judge Burns pounded the gravel for order. "Silence," he ordered.

Gene realized what he said, but it was too late to take it back.

Phil was looking down and holding onto the witness stand, exhausted. Adam and Ross began to hug each other. Allen slumped back in his chair, staring at Gene. Gene started to ramble on. He looked alternately from Judge Burns to Phil.

"I didn't go there to hurt anyone. I don't even know why the gun went off."

"The witnesses claimed you were holding the barrel of the gun when it went off."

"What? I never touched it. I was trying to push George off of me. We were supposed to play like we were fighting but he started hitting me real hard and I wanted him off of me."

It was Phil's turn to be confused. He was certain Gene had caused the gun to go off. But if the whole thing were staged by George, then what went wrong?

"Mr. Abrams, if you want to avoid a long jail sentence, you better explain what happened that night."

With the prospect of a long jail sentence in front of him, Gene was visibly shaken. His voice cracked as he began. "I wanted to start my own motorcycle business and needed a lot of money, so I placed a few bets with George. It sort of got out of control, and he threatened to have me killed unless I agreed to help him."

Phil was fascinated. "Go on."

"George told me about the fight he had with Ross and how much I looked like him. He came up with a plan to get even with Ross and get rid of his biggest rival at the same time."

"That would be Mark Dunlop?"

"Yes... George hated him. He offered to sell Bill Kennedy whatever he needed if he could get invited to the party as well. He said he was a real fan of some of the guests that would be there, and Kennedy bought all of it."

"He set me up so that I would pass as Ross at the party. He gave some coke to Kennedy to get him stoned and out of the way. Then I was to act like Ross and try and stop the sale. George and I would play like we were fighting, and George would ask Dunlop to help him. Dunlop was supposed to be pulled into the fight, and George would kill him and I would run away like I did it. Everyone would think Ross did it and George would have taken care of both of them."

A pin drop could have been heard in the courtroom as Gene continued.

"Something went wrong. I heard the gun go off and George fell on top of me. I sort of panicked and pushed him off of me and ran down the hall to get away."

"When you ran down the hall and into the other bedroom, where was Ross Daniels."

Gene didn't seem to know what to say, but he answered.

"He was coming out of the bathroom. I hit him on the head before he could see me and put him on the bed."

"You carried him to the bed?"

"Uh... yes."

"That's when the blood on your shirt got onto his?"

Gene was still somewhat disoriented. "What?"

"Mr. Abrams, how did you know Ross would be in that particular bedroom? There are a dozen bedrooms upstairs."

"One of the girls at the party worked for George. She put something in Ross' drink to make him sleepy and then pretended to help him up the stairs to sleep it off. She was supposed to leave him in that one so I could get away through the window without being seen."

Phil shook his head in amazement at the overall detail in the planning of the murder. "How did the gun go off?"

"I don't know... honestly."

There was no response from Sherlock or Allen. Phil approached the bench. "If we ever find Mark Dunlop, I believe we'll find out how that gun went off, Your Honor."

Judge Burns nodded.

"I have no more questions for this witness."

Gene stared blankly at Phil as if waiting for more questions. Mike Blackmon stood up.

"In light of this testimony, Your Honor, the State will move to drop all charges against Ross Daniels, and we request Mr. Abrams be held in custody until accessory charges can be filed against him in the murder of George Artoles."

"Mr. Abrams is remanded into the custody of the State until further notice. All charges against Mr. Daniels are hereby dropped. You are free to go, Mr. Daniels." He banged the gavel. "Case dismissed."

The bailiff stood up. "All rise."

The media made a mad dash for the exits as Judge Burns left. The two Marshalls quickly escorted a still confused Gene Abrams out of the courtroom. Adam was so grateful, he hugged Allen and the rest of the defense team. Phil braced himself and left to give a short statement to the press as Alicia hugged Ross.

CHAPTER 26

Alicia was waiting for Allen in the jury deliberation room. She ran to him and hugged and kissed him. Allen started to caress her, but she pulled away and started blurting out her relief at the trial's end.

"I just can't believe this nightmare is finally over... after all these months. Dad said Ross can pick up his things later today. And we couldn't have done it without you, Allen."

"Everyone helped on this. It wasn't just me."

"Bullshit, Allen. Why can't you take a compliment?"

"I just think it was a team effort."

"Oh, really? Well, I'm not going to do this to the team."

She wrapped her arms around him and practically knocked him backwards as she kissed him. The door opened quietly and Janet stopped when she saw them. She watched them for a moment but didn't interrupt. When Alicia stopped and let go of Allen, he saw Janet at the door.

"Janet!"

Alicia saw Janet, and stepped back a little from Allen.

"Hi, Allen. I hate to bother you. I heard the trial is over already. You must have hit a home run with your lie-detector system."

"Janet, this is Alicia Daniels, Ross' sister. Alicia, this is Janet Turner."

Alicia didn't seem to be bothered by Janet's sudden appearance.

"Hi. Pleased to meet you. I guess I better leave you two alone. I'll see you later, Allen."

She smiled as she passed Janet on the way out. Janet walked over to Allen and gave him a "polite" kiss.

"You couldn't have known about the outcome of the trial."

"No. I had planned to come out here a week ago."

"About that kiss...."

Janet laughed. "If you had just helped get my brother out of a murder charge, I would have kissed you too."

"What's going on, Janet?"

"I'd like to talk about us."

"Us?" Allen sat down warily, wondering where this would lead.

"Yes. Last week my ex-husband, Bruce, came back into town. For several months, he has been bumming off some friends who were living in Spain. He came to tell me he won third place in 'EL GORDO', the big lottery there."

"Wow! Is that a lot of money?"

"It will be if he stays in Spain and doesn't pay the U.S. taxes on it. He probably is set for life." She paused before continuing. "He wants me back, Allen."

"The fact that you're here means you're going back with him. You could have told me this on the phone the next time we talked."

"I had hoped I was over him when I started dating you, but when I saw him... I knew I wasn't." Janet sat down next to Allen and put her hand on his. "I never got around to telling you why we split up. We were constantly fighting over money. He's an artist and a dreamer, but most of all a big spendthrift. We were constantly being hounded by bill collectors, and I just couldn't take it anymore."

"And now that he's rich...."

"It's not an issue now." She seemed embarrassed. "You aren't too mad at me, are you, Allen? I would still like for us to be friends."

Allen didn't answer immediately. After a moment he shook his head.

"To be truthful, you would probably be better off with him in the long run. I'm not the ideal husband type."

"Don't let the Marys of the world convince you of that, Allen. I'm sure you would make a good husband. Who knows, if this hadn't happened, I might have had the chance to find out." She squeezed his hand.

"I really do wish you well, Janet."

"I know. And I hope everything works out for you. This latest invention of yours will probably make you wealthy some day."

She kissed him and then started to walk out the door. She stopped and looked back. "You know, Allen, you're not very good at calculating odds. You did find a blond nymph out here."

She walked out the door, and Allen slumped down in his chair. He suddenly remembered Alicia and ran to the door. He was expecting to find her nearby and was disappointed that she was gone. He sighed and walked back into the courtroom to take the system apart and prepare it for shipment.

As he began taking it apart, he heard a woman squeal in delight. His curiosity got the better of him and he opened the courtroom door to the outside hallway and peeked out. He took a short breath when he saw Alicia hugging a tall handsome man about her age. At almost the same moment, Alicia saw Allen and yelled to him to come over. Allen walked warily to them and as he neared Alicia called to him excitedly.

"Allen, the most wonderful thing has just happened. John has asked me to marry him. Isn't that wonderful?"

Allen was confused to say the least and it must have shown on his face. Alicia whispered something into her fiancé's ear.

"Okay. I'll be back in a minute." He walked off and Alicia held an engagement ring with a huge diamond up for Allen to see.

"Isn't it magnificent, Allen? John is so wonderful!"

"It's nice. Err... Alicia?"

"Yes, Allen?"

"I thought...." Allen's voice trailed off. Maybe he had jumped to some premature conclusions. They'd had some really great times together, but the questions she had asked him about Janet popped back into his head. They had not even talked about a long-term relationship. He hadn't asked her for an exclusive relationship, and he certainly hadn't told her he loved her. As he rolled the times they had spent together around his head, he couldn't remember anything she had said that indicated she wanted a more serious relationship. Maybe it had all been in his head.

"What, Allen?"

"Oh, nothing. I really wish you all the best."

"Thanks, Allen. I think you're a swell guy. I hope you and Janet can get it on like John and I will."

Allen could see John returning when he looked beyond Alicia. Allen shook their hands and wished them well. He watched them walk off arm in arm and sighed. He returned to the courtroom to finish packing the system components.

Allen had almost finished packing everything away when Ann rushed into the courtroom looking for him. She grabbed his arm excitedly.

"Allen, they have Dunlop in custody!"

Allen almost dropped the equipment case he was holding. "What happened? I thought he would be out of the country by now."

"He turned himself in. Word must have gotten back to George's father that Dunlop pulled the trigger. He said two men were chasing him and he decided his only chance was to turn himself in, so he drove directly to the nearest police station. He's telling everything in exchange for a reduced sentence."

Allen stared off into space, stunned at the sudden turn of events. "What really happened? Did he say?"

"When Dunlop went upstairs with George and Bill, he recognized several of George's men standing at the top of the stairs and became suspicious. When Gene entered as Ross and started a fight with George and George asked him for help, he guessed it was a setup. He grabbed the gun, pointed it at George and forced him to pull the trigger."

Allen shook his head in disbelief.

Ann tugged at his sleeve. "Remember the one hundred thousand dollars found in Ross' car?"

"Yes."

"That was Dunlop's. He had just come from an even bigger deal and had all that cash. Right after George was shot, he hoped they would pin everything on Ross, so he put the money in Ross' car almost under the police noses."

"How did he know which car was Ross'?"

"His license plate said ROSS D."

Allen laughed. "I missed that. I wonder if it would have helped Sherlock?"

"It doesn't matter now. It's all over."

She was so happy she hugged Allen. To her surprise he hugged her back.

CHAPTER 27

Adam Daniels was speaking with a crowd of reporters on the steps of the courthouse as Phil, Allen, Ann, and Joanne stood near the doors to the courthouse, discussing the case with Detective Johnson.

Ann remembered her interview with Gene. "What will happen to Abrams?"

"He gave us a full confession. He probably will be charged as an accessory, but we are reviewing the whole case. We're really glad we got our hands on Dunlop. We can finally sort everything out now. Abrams could just as well have been the one lying on a morgue table. He's claiming innocence, but he did go there to help George kill Dunlop, so we'll just have to see what Mike Blackmon wants to do."

"We owe you and Judge Burns that special report on the effect the Sherlock software had on the trial," commented Allen.

"I'm sure it helped, but I think some intuitive problem solving on your parts played a bigger role. I really am grateful for the help. The case against Ross was circumstantial, and really, only the bloodstain on his shirt made him the prime suspect. Oh, well, I have to go testify in another case. It's been a pleasure working with you."

Johnson shook their hands and walked off. Adam Daniels finished with the reporters and walked over to them to shake their hands.

"Phil, I can't begin to tell you how grateful I am for your help."

"I think we all owe Allen a great deal. I don't think I could have contributed much without him. After all, the Sherlock software, the image transformation of Abram's face, and the lie-detector system were all Allen's ideas."

"But ideas aren't valuable unless they're tested and validated. I would probably still be tinkering with the first prototype if Phil hadn't found a need for it."

Adam wouldn't let them downgrade their efforts. "Nevertheless, I want to do something. So... Allen, a little while ago I called and talked to Tom Kelley, and he agreed to sell me his half of Sleuth Software. I have also asked my lawyers to look at taking the company public. They estimate your stock will be worth about twenty-five million."

Phil coughed. "Whoa, Allen! How about a loan?"

Everyone laughed.

"You don't need to do this, Adam."

"My advisors think this could make me a great deal of money also, so it isn't all gratitude. It's good business as well. Ann, I have also asked them to look at setting up a joint venture with Stevens Equipment Company to exploit your ideas for a medical monitor utilizing part of Allen's system, and another one to help you and Allen in fully developing the lie-detector application, so you will do pretty well also."

"And lastly, I am offering part ownership and the job of Vice President - Legal Affairs on all of these companies to my old friend, Phil. If you can stand to leave the prosecuting business, of course."

Phil grinned. "That sounds like an offer I can't refuse."

"You all will be hearing from me soon about this, but right now I am going to see about clearing the paperwork on Ross. Thanks again, everyone."

He shook their hands again and walked down the courthouse steps toward the police station.

"I told you; Adam knows how to repay a debt." Phil reminded them as he put his arm around Joanne. "We have some celebrating to do, honey."

She turned and hugged him. "We have a lot of plans to make, Phil."

They started kissing passionately and Allen and Ann drifted slowly away to leave them alone. A large crowd of reporters began following Adam and the rest of the defense team, trying desperately to find a sound bite for the evening news. Allen was left standing with Ann on the courtroom steps. They watched Phil wave down a taxicab. As the taxi drove off, they could see Joanne and Phil intertwined in the back seat celebrating their relationship.

"Well, it was a pretty exciting day," commented Ann softly. "You sure made out like a bandit with that new company Daniels is going to start for you. How many millions did he say you would be worth?"

Allen chuckled. "Not as many as you'll have when he kicks off those new joint ventures with you."

An awkward silence passed before Ann turned to him. "You know, Allen, I have never had a better business week. So how come I feel so lousy?"

"I don't know. Why don't we go and have a drink and talk about it?"

"Where's the nearest bar?"

"I haven't checked out of the hotel yet. Why don't we head that way? There's a nice bar just off the main lobby."

"Let's go." She took Allen's arm, and they strolled off toward the bar.

The bartender delivered their drinks, and Ann held her glass out in a salute. Allen touched his glass to hers and they sipped their drinks slowly.

"I know what it is," she said finally. "I have to go home and face the fact the kids don't have a dad anymore, and I don't have a husband. It's going to take some getting used to."

Allen laughed a little. "You're pretty, rich, and successful. It won't take you long to find someone."

Ann was sipping on her drink and almost choked. She had a surprised look when she turned to him. "Do you really think I'm pretty, Allen?"

Allen had let that slip out, and he blushed. "Err... of course I do. I guess I've always thought you were pretty."

She was staring into his eyes. "Why didn't you ever say so?"

He stared back at her with a blank expression. "I don't know. I guess I thought you would think I was trying to come on to you."

She chuckled. "I always wondered why you didn't."

Allen face was flushed. "I guess I thought it might mess up our friendship."

"Allen, sometimes you think too much."

Allen laughed so hard, he almost fell off his bar stool.

Ann wondered why he thought that was funny. "We were best friends in high school, and when we went away to college I used to wait up at night for your call."

"You did?" he asked incredulously.

She nodded. "I was hoping one time you would say you missed me and wished we could see each other every now and then."

"But I did wish that! I just didn't have any money to go and visit you."

She took another drink. "I would have given you the money, if you had just asked."

Allen was thinking about her revelations when she put her hand on his. "I finally gave up on you when Phil started calling and coming to see me."

Allen was wondering how differently his life would have been if he had just picked up on the signals, she obviously had been giving him and he totally missed. She was smiling when he finally looked at her.

"It looks like we missed some opportunities."

She moved closer to him and they kissed.

"Allen, didn't you say you haven't checked out yet?" She closed her eyes as Allen held her against him.

CHAPTER 28

The afternoon sun was setting, and a gentle breeze was blowing in through the open balcony door, but Allen and Ann were totally oblivious to the outside world. Ann's hair had fallen across his face, and Allen could smell the coconut oil in her shampoo. She lifted her face to look at him.

"I wish the world would go away and we could stay here just like this."

Allen seemed apologetic. "Ann... this IS fantastic, but I need to go to the bathroom."

She laughed loudly and rolled off him. "Just hurry back."

Allen did hurry into the bathroom. When he came out, he saw Ann standing with her arms folded by the balcony door, watching the setting sun. Allen tried not to, but he found himself comparing her to Alicia. Being around Alicia had been intoxicating, but somehow, the thought of a new intimate relationship with an old friend like Ann was even more exciting. He knew Ann worked hard at physical fitness and he stood admiring her body for a few moments. The afternoon breeze was blowing her hair about as he walked up behind her, hugged her, and started kissing her neck.

"I love you, Ann," he whispered. "I think I've always loved you."

She turned and put her arms around him. "I didn't think I would ever hear you say that" she said as she gazed into his eyes. "I love you too, Allen."

They stood hugging each other for a while.

"What about Tim and Susan?"

"They've known you all their lives as 'Uncle Allen', it won't take them long to call you 'dad.'"

He held her tightly as she kissed him.

"Partners again?" she asked.

"In every way."

They kissed until Allen pulled away.

"Uh... Ann?"

"Yes?"

"About those missed opportunities...."

"Yes?"

"Do you think we could catch up some more?"

Allen grinned and Ann put her arms around his neck.

"All right. I'm willing... if you're able."

"Trust me... I'm able."

Before they could resume their "catching up," the doorbell rang.

"Oh, shit!" Ann exclaimed and ran into the bathroom.

Allen grabbed a bathrobe, slipped into it quickly and answered the door. Three large men in dark business suits were standing outside. From their look and demeanor, Allen sensed they were not the media or in any way related to the trial just concluded. The word "military" flashed through his mind.

"Yes?"

A tall man with a ruddy complexion and a blond crew cut flashed his badge in Allen's face.

"Dr. Allen Atkins?"

"Yes?"

"I'm John Reems with the Central Intelligence Agency. May we come in?"

Allen hesitated. CIA? What if the badge was a fake? What if they were LoBlanco's or Antonio Artoles' men seeking revenge? "Well, I'm kind of busy right now. Could you come back in a little while?"

"Actually, the matter we would like to discuss with you is rather urgent. We would like to bring you and Mrs. Stevens to our office to discuss a matter of national security."

Allen was more than a little concerned that they even knew Ann was with him.

"Just a moment."

Allen closed the door and reflected on his options. He quickly picked up Ann's clothes and knocked on the bathroom door. Ann had a bath towel wrapped around her when she opened the door. She saw Allen's concerned expression.

"What is it, Allen?"

"Three men claiming to be CIA agents are here. They even know you're here and want to talk to us at their office."

"CIA?" Ann stared at him in disbelief until the doorbell rang again. Allen handed her clothes to her. She closed the door to dress.

"JUST A MOMENT!" Allen dressed hurriedly and opened the door. As Reems entered, Allen saw the other two men take up guarding positions outside the door. He heard the bathroom door open. Reems brushed past him and showed his badge to Ann.

"Mrs. Stevens? My name is John Reems, with the Central Intelligence Agency. We need to talk to you and Dr. Atkins."

"What's this all about?"

"A matter of national security. We would like to talk to both of you at our office here in Los Angeles."

"Why would you want to talk to me?"

"I can't discuss the details here." Reems seemed in a big hurry to leave.

"What if we don't want to go there?" Allen was still suspicious.

"I'm afraid you have no choice. If you resist, we'll have to have this discussion at the police station."

Ann appeared frightened, but Allen was visibly relieved. He explained his concern about a possible visit from LoBlanco or Artoles.

Reems laughed. "You won't have to worry about those two anymore. The FBI arrested both of them earlier today."

Allen must not have understood him. "FBI?"

"RICO issues." Reems opened the hallway door. "Let's go, please."

Allen took Ann's arm, and they went with Reems down to the lobby and out of the hotel. The other men surveyed the area around the hotel entrance and signaled them to enter a black limousine waiting outside.

CHAPTER 29

Reems had not given a hint of what was going on during the ride to the CIA office. Allen put his hand on Ann's to reassure her. When he felt how cold her hands were, he put his arm around her. Reems was expressionless. When the car stopped, several agents were waiting to escort them. It was already dark as Allen and Ann followed Reems into a large commercial building in the downtown area of Los Angeles. They took an elevator to the 25th floor. Reems swiped a security badge unlocking one of a pair of large glass doors emblazoned with «Carborundum, Inspection and Analysis» in large block letters. He felt a little relieved when they entered a small office and he saw Reem's name on a desk nameplate. He doubted LoBlanco or Artoles› men would have displayed their names in that manner.

Reems' office was furnished with standard government-issue furniture. Unlike some other offices they had passed, his office was remarkably clean and free of paperwork. There were two large, comfortable chairs in front of his desk, which was also devoid of paper except for a manila folder. A large black executive chair was waiting for Reems.

"Please have a seat. Would you like some coffee?"

Allen declined but Ann accepted his offer. Reems spoke softly to an agent standing outside his office and closed the door.

"I'm sorry we couldn't do this tomorrow, but we understand both of you were scheduled to return to Houston in the afternoon."

"What exactly ARE we doing, Mr. Reems?" Allen's voice betrayed his ire at their forced attendance and the lack of information supplied so far.

Reems sat down behind his desk, opened the folder, glanced at it, and continued.

"The CIA was routinely informed when Mr. Adam Daniels made a request to buy or borrow certain classified surveillance equipment. We reviewed the request and noted, with some interest, the proposed use involved integrating the equipment into a newly developed lie-detector system."

As Allen and Ann glanced at each other in surprise, the door opened, and an agent handed Ann a cup of coffee. He closed the door behind himself as he left, and Reems continued.

"We approved the request, and several of our agents attended Ross Daniels' trial to view the operation of the system," Reems chuckled. "You probably didn't see our agents, Dr. Atkins, but they were watching the operation of your system, practically over your shoulder."

"I did see them. I just thought they were just a bunch of nosy reporters." Ann commented. "They really were... sort of."

Reems continued. "We were very impressed at the operation of the system and the ease with which it allowed Mr. Conley to revise his questioning of the witnesses based on their answers. We were pleased that our equipment could be integrated into your system and help solve a murder case."

Allen interrupted him. "That's fine, but why would any of this be of interest to the CIA?"

Reems sat back in his chair. "Now that your system has been proven, we have an urgent need for it. As I mentioned earlier, a matter of national security."

Ann sat her coffee cup down on a nearby table. "What does this have to do with me? I know very little about the system, other than its physical size requirements."

"We know you helped package the equipment and know quite a bit about its mechanical operation. We may need to make some additional modifications, and you would be in an excellent position to help us."

"I have a feeling this is more than a request," Allen said, looking at Ann.

"This is an urgent matter, Dr. Atkins. It simply can't wait. I'm afraid we are going to need both of your services for the next month."

Ann was trying to control her temper. "Month? I have two small children I haven't seen in several weeks. I don't have time for this! I'm sure you can make whatever changes to the system you need... without me." She looked at Allen. "I'm sorry, Allen, but Tim and Susan need me. I would like to help... and be with you, but my mother can't take care of them forever."

Reems answered before Allen could.

"For security reasons, we already have stationed a team of agents with them, including several women agents who can help your mother take care of them. We'll also make sure you stay in frequent contact with them and...."

Allen interrupted. "Is this really necessary? I'm sure I can help your technicians make whatever modifications are required."

"Sorry, but we can't waste any time. It's not negotiable."

"What exactly does that mean? Are you threatening us?"

"I didn't mean to make it sound like that, Dr. Atkins. I just mean we have already addressed those potential concerns. We really do need your help."

He paused and turned away for a moment, as if preparing an argument asking for their help.

"People are asked to help their nation in many ways. You happen to have developed a tool we have an urgent need for. There isn't time for you to train our people on its use, so we must ask for your help."

"We both have jobs. I own a small company and can't afford to be away from it for a month," commented Ann. "People will wonder what happened to us."

"We can brief NASA management in regard to Dr. Atkins. You will have to determine the best way to inform your managers of your absence, Mrs. Stevens." He made a few notes in the file folder. "We'll also make certain you stay in touch with them by phone and e-mail."

Allen and Ann sensed the futility in trying to resist any further. They obviously had planned this for some time.

"All right," Allen acquiesced. He looked for her support. "Ann?"

She stared at him for a moment. She still had some misgivings, but at least she would be with him. "Okay. But... now what? You still haven't told us what this is all about."

"You will be briefed." Reems seemed relieved as he glanced at his watch. "It's time to get going now."

Ann tried one more time. "Where are we going?"

Reems smiled at them. "That's classified also. We have an airplane waiting for you, and time is of the essence."

"What about our clothes?"

"A team has collected all your belongings and taken them to the airport. We also collected all the lie-detector system equipment as well."

"I almost feel like we are being kidnapped."

"We don't want you to feel like that, Ann, but the urgency of the matter requires us to act quickly."

Allen suddenly remembered Phil. "What about Phil Conley? Is he involved?"

"We would like to utilize his services as well, but so far, we can't find him or his wife. They didn't leave a forwarding address. We found their address books and tried every number in them, but none of their friends seemed to know where they are. Do you happen to know?"

Allen glanced at Ann. "They're on a long honeymoon. Phil wouldn't tell me where they were going though." Ann shrugged her shoulders when Reems looked at her.

Reems stood up and began to herd them out the door. "We'll find them. But, for now, you need to go."

"You? Aren't you going as well?"

"I would like to, but it's beyond my jurisdiction."

"Where did you say it was?"

Reems smiled at Ann. "I didn't say."

He locked his office and escorted them back to the waiting limousine.

"Have a safe journey," he said as he closed the door.

Allen and Ann hugged as the limousine pulled away from the curb and sped off into the night.

CHAPTER 30

Allen watched several signs for the Los Angeles Airport pass by before the limousine pulled into a private, fenced-off section and stopped at a guard gate. The driver flashed his badge, and the guard opened the gate. They stopped near a large business jet and the driver opened the limousine door.

Allen suddenly sat upright in his seat. "A Citation X+!" It seemed like an ordinary business jet to Ann, and she was amused when Allen almost ran over to the jet. Two well-armed agents with partially concealed machine guns were also watching Allen's every move. Ann smiled as Allen walked along the fuselage feeling it with his hand as circled the jet.

He's almost caressing it. She thought.

When he walked back to her, he was almost bubbling over. "This is the fastest business jet in the world. It has a ceiling of 51,000 feet and flies at just over 700 miles per hour, almost Mach 1!"

Allen continued to rattle off the plane's specifications, until he saw Ann yawn. "How could you possibly know all that?"

"I have a friend that sells and leases business jets like this one. I never thought I would get to fly on one, though."

"I just wish we knew where we are going." Ann saw him frown. "What it is, Allen?"

"This jet has a range of about 3800 miles," Allen commented as he watched the plane being fueled.

"So?"

"That limits its range to anywhere in the U.S. or Alaska, or Hawaii, maybe. It can't make it across the Pacific without refueling. Maybe we can tell where we're going if it makes a refueling stop."

"I wonder why we aren't going on a regular commercial flight," Ann mused out loud.

"I would guess they don't want a record of our trip." Allen made the comment in an offhand, absent manner, but it triggered an alarm in Ann.

Why not?

Allen and Ann watched as their suitcases and the aluminum suitcases containing the lie-detector system were loaded into the cargo hold.

"Well, at least we have our clothes," Ann said with some relief.

The co-pilot emerged and motioned them to enter the jet. Allen squeezed Ann's hand and they climbed the ladder and entered the jet. Two men and a woman were already on the plane as they entered the eight-seat main cabin. None of them appeared to be eager to talk about themselves or why they were on the flight. Allen and Ann settled in for a flight of undetermined length. Allen noted that they left precisely at midnight.

Almost like a scheduled flight.

The hours passed slowly. Allen tried to sleep but couldn't. Ann had finally dozed off while cradled on Allen's shoulder. He watched her sleep and marveled at the renewal of their relationship, one that had been dormant since high school. He kept running the events that had occurred since the trial over in his mind. What was so urgent? Why did the CIA need the lie-detector system? There were some pieces of the puzzle missing and Allen finally decided that, without them, he wouldn't be able to put the picture together. He eventually dozed off.

The steady drone of the plane's engines slowed, and the jet began a rapid descent, waking Allen. It was still dark, but he could see the moon's reflection off the ocean as they neared an island's runway. The captain made a brief announcement and the plane leveled out from a steep banking turn, touching down gently on the runway. Allen looked at his watch. It was 5:47AM in Los Angeles but it was still nighttime here.

Allen strained to pick out details in the dark but couldn't recognize anything. He had been dozing and missed the first few words of the announcement. Ann stirred and opened her eyes. When she saw Allen, she smiled.

"If this is a nightmare, I'm glad you're here."

Allen kissed her. "I can't tell where we are. I don't recognize anything."

Ann stared out the window. "Some military base."

"We must be heading almost due West."

Ann yawned. "How come?"

"If we had headed North or South the sky would be getting brighter. If we had headed East, it would be the middle of the morning."

The plane stopped, and the co-pilot opened the cabin door and addressed the passengers. "You can have a few minutes to stretch while we refuel."

Some of them stood up to leave. Allen and Ann followed the co-pilot down the jet's ladder. At first Allen didn't notice anything out of the ordinary. When he paid more attention, he saw several almost concealed weapons on the maintenance personnel refueling and servicing the jet. The passengers milled around near the ladder. None appeared eager to venture off to the nearby building and hangers.

Allen watched the jets refueling. "Another 3800 miles?"

Ann watched the service personnel load a significant quantity of food and drinks on the plane. "That's a lot of food. Where could we be going?"

Allen looked at his watch. "I would guess we are somewhere in the middle of the Pacific, heading for Asia."

"Japan?"

"Probably not. They could have put us on a one-stop commercial plane to Tokyo."

"Where then?"

"I would guess a small, private or military base in Thailand or South Korea."

Allen pulled a pen from his coat pocket and a scrap of paper from another pocket and began making some calculations.

"Based on the average speed of this jet and the elapsed time, I would say we are only about one-third of the way there, assuming it's somewhere in Southeast Asia."

The jet did make another refueling stop, and without a word the passengers exited the jet to stretch. The jet had stopped at the opposite end of the landing field from the control tower and taxied to a remote building just off the runway. Ann tapped Allen on the arm and pointed to a small convoy of trucks and vehicles heading in their direction.

"Welcoming committee?"

Allen shrugged his shoulders. "I doubt it. They look more like maintenance vehicles."

Several service trucks surrounded the jet as it was refueled, food trays and drinks transferred, and the oxygen supply replenished. Allen glanced at his watch and scribbled "12:30PM in LA" on his "back of the envelope" calculations. The sky was just beginning to lighten, and Allen guessed the local time to be about 5:30 in the morning.

Ann peeked over his shoulder as he tried to determine where they were. "Where are we, Allen?"

"I'm not sure. The military has numerous airstrips in the Pacific."

"Where now?"

"Who knows? Almost all Southeast Asia is within range from here."

Ann looked about questioningly. "Where is everyone?"

The other passengers had disappeared.

"Maybe in those cars."

Allen and Ann watched the small convoy gather on the taxi strip and head back to the control tower and hangers.

The co-pilot motioned them to re-board. The flight crew made no announcements and the jet taxied to the end of the runway. Ann and Allen were slammed back into their seats as the jet hurtled down the runway and lifted off into the early morning light.

Sheer boredom had set in, and Allen and Ann had finally managed to fall asleep when the jet made a rapid descent into a large airport near the outskirts of a major metropolitan area. Allen jerked awake as the jet touched down abruptly and braked hard to stop at a building housing an aviation service company. He guessed the local time to be the late morning as he strained to make out the surroundings. The jet's door opened and two armed personnel wearing maintenance coveralls entered. The co-pilot touched Ann on the arm to awaken her. She saw the weapons and exchanged a worried expression with Allen. He tried to reassure her.

"I'm sure everything is all right. We'll do this as quickly as possible and be heading home in no time at all."

"Reems said a month, Allen."

"He just didn't realize how efficient we are."

He tried to lighten her spirit by smiling at her, but her expression reflected her growing concern that this adventure was not going to be over quickly.

The two security agents suddenly stepped back and a tall, burly man in a nicely tailored suit crowded past them to introduce himself.

"Good morning, Dr. Atkins, Mrs. Stevens. Welcome to Bangkok!"

"Bangkok!" Ann and Allen replied together.

The agent frowned at their response. "Yes... my name is John Dough, your contact here."

Ann laughed loudly. Of course, it is."

Dough seemed prepared for this response and handed them a business card, identifying him as Regional Sales Manager for Carborundum.

"Oh, sorry," said Ann when she noticed the spelling of his last name.

"That's all right. It's only a problem here with Americans... or movie fans." He sat down on a seat near them. "I'd like to bring you up to date on the current status."

Allen pulled a soft drink out of a nearby beverage cart and opened it. "Status of what?"

Dough seemed confused. "Didn't Reems brief you on the nature of your assistance."

"NO!" Ann and Allen replied together.

"Well... Reems IS a little cautious, but this isn't the right time or place either. A limousine is waiting outside to take you to a hotel. I'll come by this evening and fill you in on why you are here and bring you up to date." He stood up. "I'm sure you are tired and would like to freshen up some. Please!" He motioned them to the door.

Compared to the cool island breezes at the refueling stops, a nearly suffocating combination of heat and humidity enveloped them as they exited the jet.

"Just like Houston in August," commented Allen to Dough.

Dough tried to suppress a laugh but couldn't. They were quickly hustled into the front limousine of a pair waiting near the jet. Dough held the door of the limousine open for them. When they had entered, he paused before he closed the door. "Please dress as tourists this evening to avoid drawing attention to yourselves." He closed the door and Ann and Allen stared at each other for a moment. Ann peered out a window as she buckled her seat belt and was relieved to see their luggage mixed in with the aluminum suitcases containing the lie-detector equipment being transported to the limousines.

CHAPTER 31

Their limousine merged into the heavy late morning traffic of Bangkok as it exited the airport, with the second following closely behind. Allen and Ann gawked at the complex mix of bicycles, three-wheeled motorized taxis, buses, regular taxis and cars as they entered an area crowded with skyscrapers and high-rise hotels. Allen and Ann didn›t notice the other limousine following them in the heavy traffic and Ann sighed with relief as they pulled into the large circular drive in front of a four star hotel.

Ann saw a sign for the Chaophya Park Hotel Park Hotel and tried to pronounce it several times before giving up. The doorman opened the door and stepped out of the way. He instinctively sensed they didn't need any further assistance as several agents quickly picked up their suitcases and carried them inside.

As they entered the hotel's massive lobby, an agent handed each a room card key and motioned them to follow him to the elevators.

As they were waiting, Ann whispered to Allen. "I really don't want to be away from you. Could I share your room, Allen?"

Allen grinned and suddenly didn't feel as tired. "Of course." Ann laughed and gave him a little push as the elevator opened.

The elevator stopped on the 15th floor and Ann started to walk with Allen to his room. An agent spoke just loud enough to get her attention. "Your room's this way, Mrs. Stevens."

Allen smiled at her when she replied, "I'd prefer to stay with my fiancé."

The agent nodded. Allen opened the door to a large, luxurious suite with a massive king size bed.

"This will do nicely," Ann commented as she ran to the bathroom.

The agents left their suitcases and walked off. Allen watched one of them sit down on a bench in the hall and start to read. He closed the room door and stood staring out the balcony door at the city wondering why they had been brought halfway around the world. Ann put her hand on his arm and shared the view for a moment.

"The shower's big enough for two," she commented without looking at him.

A huge delivery truck lumbered down a broad boulevard lined with palm trees and turned onto a narrow street that wound lazily through a quiet suburban neighborhood in the northern part of Bangkok. It stopped in front of a large house surrounded by a white stone fence. Two men unloaded two large cardboard boxes containing a washer and a dryer. They were greeted by a conservatively dressed woman who opened a gate to the driveway. The boxes were quickly brought into the house through a side entrance near the garage. A short while later the same men brought out pieces of the cardboard boxes, along with an old washer and dryer. They loaded them onto the truck and drove off.

At precisely six in the evening, Dough phoned to inform them he was on his way to their room. A few minutes later, he entered with a steward carrying a large pot of coffee. The steward gathered the remnants of a prior room service delivery and left quietly. Like Ann and Allen, Dough was very casually dressed, indistinguishable from a tourist, and carrying a large folder. "How's the room? Did you get a chance to rest."

"Yes," replied Ann. Several times.

"Excellent!"

They sat down at a small table in the room and Dough pulled some papers out.

"Approximately three days ago, a North Korean entered our embassy in Cambodia. He claimed to be the head of their nuclear weapons development program and offered to exchange information in exchange for asylum. Unfortunately, he doesn't match our description of that person. We have interrogated him almost continuously, but he refuses to cooperate without a guarantee of asylum."

"Why don't you give him asylum?" Allen wondered out loud as he watched Dough pour three cups of coffee.

"We tried, but for some reason he wants to speak to a member of the State Department first. It seems he doesn't trust us." Dough remembered an internal joke and chuckled.

Ann frowned at him. "I hate to sound naïve, but why don't you let him contact them?"

"We don't get the State Department involved in cases like this, for several reasons I can't explain."

Allen folded his arms in frustration. "So, why are we here?"

"We have tried the usual lie-detector methods and he fails them, but there are some elements of truth in his statements as well. We hope your system can help us ferret out the whole truth."

"What if he's telling the truth?" Ann sipped on the hot coffee and grimaced at the poor quality and burnt taste.

"Then there are some serious issues for the South Koreans to address. We must be sure, though, before we tell them about him."

Ann put the cup down. "Is he here in the hotel?"

Dough laughed. "No… he's in a safe house about an hour's drive from here. The team assigned to him has been briefed about the nature of your assistance and status here and is waiting for you. You'll be greeted by a woman whom the neighbors know as Mrs. French. You will introduce yourselves as Mr. and Mrs. Rose."

Allen suddenly laughed. "Is there a password as well?"

Dough was not amused. "No… but as a matter of fact, you will say you are there to discuss some church-related matters."

Something Dough said, or the manner that he said it, brought home the seriousness of what they were about to embark on. Ann cleared her throat. "When do we start?"

Dough closed his folder. "We need to leave now, if possible."

As they stood up to leave and Allen waited for Ann to get her purse, he asked Dough, "What about the lie-detector equipment?"

"It was delivered this afternoon."

Allen and Ann were waiting with Dough in front of the hotel for a limousine and didn't notice a taxi that stopped in the drive near them until Dough opened the taxi's door. "This way, please." He noticed their surprised expressions. "It's much less conspicuous."

Ann and Allen entered, and Dough leaned over to speak to them through an open window.

"Don't do anything to draw attention to yourselves. Act like you're going to visit an old friend. Kim will make sure you arrive safely."

The Thai driver turned and smiled at them. "We have air conditioning, if you like."

The evening air was considerably cooler than the stifling heat they had encountered earlier in the day, but Allen quickly rolled his window up. Dough stood upright and waved to them as they drove off. Allen was glad he wasn't driving as the taxi wove in and out of a mass of vehicles and humanity. Pedestrians reluctantly parted to the repeated honks of the driver's car horn. Soon they were out of the main part of the city and sailing along a wide boulevard lined with palms. The driver made a quick left turn and a daring dash through the oncoming traffic onto a narrow road that wound through a moderately wealthy neighborhood. Ann was commenting on the immaculate lawns and well-kept homes as the taxi suddenly ground to a halt in front of a large house surrounded by a tall, white stone fence. The driver jumped out, glanced up and down the street, and seemed to be satisfied no one was following them. He opened the door for Ann.

"This is it, madam, sir."

"Thank you," replied Allen as they got out. "Uh... I'm sorry I don't have any Thai money to give you." Allen reached for his wallet.

The driver held up his hand. "Oh... no! Mr. Dough has taken care of this. Please!" He motioned them toward the house. Allen felt strange as they walked up a short set of white stairs to the front door. He wasn't carrying anything and wondered if it didn't seem odd for visitors to arrive without a gift or something for the host.

Ann noticed several small children playing in the front yard next door. Several adults were standing nearby watching them play. Probably their parents. Down the street several other children were playing on the sidewalk. No one seemed to be paying attention to them. It seemed to her to be a typical suburban scene.

Hak Suhendra was becoming desperate to seal the potential security leak posed by the defection of Kim Sang. As section head of ⟨internal affairs›, he had already authorized two separate strike teams to ensure Sang did not reveal the wealth of knowledge Suhendra was certain he possessed.

His own superiors were becoming more and more strident in their demands that he provide a quick end to the situation. Luckily, they didn't know how much the previous unsuccessful attempts had already cost the North Korean intelligence agency.

By his own estimation, time had almost run out, and Suhendra was forced to seek the assistance of a former member of the Syrian secret police now operating as a freelance agent. He had almost choked at the price mentioned by the contact that represented Mahmoud Raman. Raman had earned his reputation during two decades of intelligence gathering. He was especially effective at counterintelligence.

Raman prided himself on his ability to travel undetected in alien territory and his world-class marksmanship. He was especially a threat to the unsuspecting soldier or security guard, as he had no qualms in eliminating those unfortunate enough to be in the way of his current assignment. As a measure of his success, he had survived repeated attempts at capture or elimination by the Israeli Mossad and U.S. intelligence agencies.

Raman had personally selected three trusted comrades for this mission. Pasha Assad was a weapons and explosives expert retired from the Syrian military with a similar number of years of service. Rasala Attar was a "scrounger," capable of finding virtually anything the team would require through an extensive network built up over a number of years spent in various espionage activities. Shim Rambaud was a transportation expert, capable of driving, flying or steering virtually any vehicle, aircraft, or nautical vessel.

The assault team had received the exact location of the safe house from Suhendra and had arrived almost at the same time as the truck that delivered the lie-detector equipment. They quickly completed a preliminary site assessment and developed a detailed plan for a night assault on the house.

Suhendra had been clear on the goal, bring Sang back alive if possible; if not, leave no witnesses. Rambaud had stolen a utilities service van and parked down the street from the safe house to allow the team to monitor its activities and estimate the potential forces they would be facing. As evening approached, a taxi passed their van and stopped at the safe house. The four mercenaries watched through binoculars as Allen and Ann walked up the stairs to the front door of the safe house. Shim Rambaud glanced at Raman.

"Agents?"

Raman lowered his binoculars and shook his head slowly. "No. Technicians, perhaps." He peered through the binoculars again for a moment and then laughed softly. "Maybe... psychiatrists."

The other members of his team laughed.

Raman looked at his watch. "It will be dark soon. If there are no reinforcements, we go in at 21 hundred hours. It is now 1830 hours."

The others synchronized their watches.

Allen rang the doorbell and a pleasant, well-dressed woman in her early 30's answered the doorbell. «Yes?»

"Good evening," Allen began. "Are you Mrs. French?"

The woman's expression didn't change, but she stepped forward and answered softly. "Yes?"

"We're the Roses and we've come to talk about some new church-related software."

"Oh, yes. Please come in." She stepped back to let them enter. "We were expecting you. Please come this way." As Allen and Ann entered, two men in jogging suits lowered their automatic rifles and sat down near the front windows to resume a round the clock vigil. The female agent closed the front door and led them from the entry foyer through a large room that served as both the dining room and living room and down a long hallway past the kitchen and a breakfast room to a large family room in the back of the house. Several men were standing in a group discussing something in a low tone when she addressed one. "John, the Roses are here."

A tall blond man about her age stepped forward to meet them. He waited until the female agent left to introduce himself.

"Dr. Atkins?" he inquired as he shook Allen's hand. "John Talbot."

Allen introduced Ann. "This is Ann Stevens."

"We appreciate your volunteering to help us, Dr Atkins, Mrs. Stevens. Your equipment is set up in the next room and I...."

"Please call us Allen and Ann," interrupted Allen.

Talbot nodded. "Oh, certainly. Please come this way."

The large family room seemed to be the focal point for the living area in the rear of the house. Almost a dozen doors opened to four or five bedrooms, two bathrooms, a study, a utility room, a library, and another room Allen guessed to have been a bedroom that had been converted to a communications center, full of various kinds of radios, TVs, and computers. A metal door led to the garage in the back of the house.

As they followed Talbot, both Ann and Allen were thinking that the former owners had done an excellent job at decorating the large, comfortable house, even before it became a safe house.

Talbot opened the door to a large utility room where they found the gleaming aluminum suitcases housing the lie-detector system lying on a large table used for folding clothes. It seemed odd to Allen that there was no washer or dryer in the utility room of the house. He also thought it odd that there were half a dozen chairs in the room.

However, there was a large refrigerator, and Allen opened it to find it full of soft drinks. He took two out and handed one to Ann. Allen began taking the system components out of the suitcases to examine them when Ann interrupted him.

"Allen, look at this."

Ann was standing near a one-way window in the room, and pointed to something beyond it. Allen put down the parts he had taken from the suitcases and stood next to Ann. Two individuals were in a heated discussion with an Oriental seated in what appeared to be a garage that had been converted into a playroom with exercise equipment and a pool table. Two security agents were stationed near the entry door from the house. Another man appeared to be monitoring some sophisticated electronic gear located on a table near the Oriental, while yet another agent stood directly behind, watching him. All the agents in the room were casually dressed as if they had just come from a sporting event. The North Korean was wearing a sweatshirt and jeans. He was sweating profusely and appeared at times confused, defiant, and generally uncooperative.

"Interrogation?" Allen wondered aloud.

Ann pointed to the equipment on the table. "Some of that equipment looks just like your stuff, Allen."

"It certainly does." Allen recognized some of the equipment. Obviously, someone had tried to put together a system somewhat similar to his.

He wondered how successful they were. Allen had been staring off into the distance, and when he glanced back into the room, he realized the interrogators were gone.

The door opened and the two men they had seen questioning the Oriental entered. The first to enter introduced himself.

"Dr. Atkins? I'm Roger Douglas, and this is Michael Gore. We're glad you're here." After shaking his hand, they introduced themselves to Ann.

"It certainly is a pleasure to finally meet both of you." He continued. "I've been following your work with great interest. As you can see, we've tried to duplicate as much of your system as possible, but there seems to be something missing. We just can't quite get it to work correctly."

Ann glanced through the window at the computer connected to the equipment. "He must be referring to Sherlock, Allen."

"Sherlock?" Douglas and Gore exchanged questioning looks.

"Yes. A proprietary software system that integrates the different measurement techniques and produces an estimate of the truth of an answer."

"And this software system is contained in this equipment?" Gore was pointing to the lie detector equipment that had accompanied them on their long journey.

"Yes."

"Excellent!" Gore started examining some of the equipment in the suitcases. Douglas was waiting for Allen or Ann to comment or ask a question.

"Mr. Douglas?" Allen said finally.

"Yes?"

"Would you mind answering a question, now that the CIA has dragged us halfway around the world?"

Douglas' face was expressionless. "I will... if I can."

"I'm sure you are probably very good at intelligence gathering. You and Mr. Gore are probably experts and knowledgeable in the many methods used to obtain information. What could this system possibly add to your considerable knowledge and limitless resources?"

Douglas unexpectedly laughed out loud. "You are too humble, Dr. Atkins. You have developed a unique system that could help us immensely." He motioned them to sit down. As they settled into the odd mix of chairs in the room, the door opened and the agent they had seen earlier near the monitoring equipment entered. Ann and Allen immediately sensed he was in charge of the operation.

Allen stood up as he moved forward, a person with a goal, and in a big hurry to complete it. He introduced himself brusquely. "Dr. Atkins, Mrs. Stevens, I'm Special Agent Joseph Harrows. Mr. Douglas and Mr. Gore are with the agency, on special assignment to me for the duration of this action."

Ann felt a sudden chill as she shook Harrows' hand. Harrows was a little taller than Allen with a powerful build, dark blue eyes, and black hair cut short in a no-nonsense style. Something in the back of her subconscious told her this was a tough, resourceful man used to perilous situations and dealing with equally dangerous and desperate men.

I'm glad he's on my side.

Harrows motioned Allen and Ann to sit down and sat in a chair opposite them. "I assume John Dough explained the nature of the problem we have here?"

"Yes. You have a potential defector and haven't been able to determine if he is who he says he is or not with your conventional techniques."

"That's correct."

Allen paused, considering the situation. "What about truth drugs?"

"If he's a double agent, they will have given him enough drugs to be resistant to almost anything we can use on him. We can't rely on drugs to be sure he's telling the truth."

"Why are we here? Why didn't you bring him to the U.S.?"

"The answer is rather complicated, as you might guess."

The answer seemed obvious to Ann. "Dough says he failed a regular lie-detector test. Why don't you tell him you don't believe him, and just let him go?"

Harrows was anxious for them to assemble the new lie detector and start the questioning, but he patiently continued to answer their questions. He needed their full cooperation. "Dough also told you that he appears to be telling the truth in some areas."

"Yes," mused Allen. "Can you tell us what he said that is true?"

Harrows nodded. "He says he was working in China and Laos and found someone willing to help him make it down the Mekong River. He said he had some close encounters with intelligence agents who were already looking for him when he crossed the border from Laos into Cambodia. He walked into our embassy in Phnom Penh."

Harrows stood up and pulled a soft drink from the refrigerator as he continued.

"We thought we had the real thing when an ops team, sent to bring him from the embassy to Thailand was ambushed by mercenaries as they were returning with him. The mercenaries seemed to be well-informed, but they were no match for the ops team, who rather easily eliminated them and brought Sang here without further incident. At this point he doesn't want to go anywhere else until he makes a deal for asylum. If he's telling the truth, the South Koreans could go ballistic on us when they hear what he has to say."

Harrows paused to finish his soda. "So, you see, we have to make sure he's genuine."

"How safe is this house, Mr. Harrows?" Ann was suddenly concerned about the possibility of another attempt on the defector's life.

"We initially had him at a less secure location in Bangkok, and they somehow found out that location before we could question him. That attack was not any more successful than the first, but we decided to move him to a more secure location. This site is very secure. You don't have anything to worry about." Harrows watched their faces for an indication of their thoughts and concerns.

"He must be genuine if they want to kill him that badly."

"That's what we thought at first, but the attacks weren't that effective, almost as if they were for show rather than to eliminate him. Just a gut instinct though. They could have just underestimated our ability to protect him. We must know if he's for real so we can get him to safety or release him."

Harrows was becoming impatient. He stood up. "How long will it take you get ready, Dr. Atkins?"

"Assembly plus checkout... about an hour Mr. Harrows."

"Fine. I'll be back then." He left quickly, leaving Ann and Allen to stare at Douglas and Gore. Gore smiled at them.

"Don't mind Harrows. He's usually a lot friendlier than that. I guess he's under a lot pressure to get this issue resolved."

Allen chuckled. "I've dealt with worse at NASA. Well, time to get to work."

He stood up and began pulling components from the aluminum suitcases. Ann, Douglas and Gore quickly joined in.

Allen had barely completed the assembly when an explosive sound reverberated through the house. The lights were quickly shut off and agents took up defensive positions. A few tense moments passed until the lights came back on and word passed around the noise was just a car backfiring as it hurried down the street. Everyone breathed a sigh of relief, especially Ann.

The lights in the living room were still off as Joseph Harrows peeked out through the curtains and watched the oily blue cloud of smoke left behind by a wreck of a car, slowly dissipate in the gathering darkness. Harrows was a cautious man. His attention to detail saved his life more than once. He stared out the front window, thinking.

Pasha Assad carefully placed a small explosive device under the trunk of an old ramshackle car he had bought for a few hundred dollars at a used car lot and painted to look like a taxi. Rasala Attar hired a taxi driver for the evening and instructed him to drive past the safe house a few times. Assad connected a radio-controlled trigger to the explosive and waited for Raman's command. When Raman's team was in position, he signaled Attar. The old car sputtered and coughed as it lurched toward the safe house. When it was in front, Assad pressed a button on the remote control and the device exploded loudly, sending a large cloud of smoke out the back of the vehicle.

That should put them off guard. Raman signaled the team to move in. Almost in unison, Assad and Raman attached suction cups to windowpanes and began tracing circles with glasscutters. They reached inside and unlocked the windows and lifted them as quietly as possible. Raman climbed in a back bedroom window and quickly stood behind the bedroom's door as Rambaud crawled through the open window. He unshouldered an automatic rifle, looked at his watch, and began a short countdown.

Allen completed a systems check and looked at his watch. «We still have a few minutes,» he commented to Ann. «I'll be back in a minute.» While Allen left to find a bathroom and something to drink, Ann settled into a large, overstuffed chair in the family room. A ceiling fan swirled lazily overhead and the gentle buzz of conversation by several agents standing near the communications room proved to be relaxing. She felt the tension that had built up slowly draining away. After a few moments, she dozed off and began to dream... a dream that soon turned into a nightmare of being pursued. Just when she felt she had found a safe place she stumbled over something. She looked back to find Allen lying in a pool of blood. She somehow knew he was dead and cried out in anguish... so loudly she woke up breathing hard and trembling. She realized it was only a dream, but it left her shaken and angry at the position the intelligence agencies had put them in.

Security Specialist Tom Smith had been startled by Ann's outcry and stood up from his monitoring station to see what was wrong. A computer flashed a silent alarm and an outline of the safe house's security system appeared on the screen. The two alarm zones flashing red went undetected for a critical moment.

At exactly 2100 hours, Raman and Rambaud pulled gas masks over their faces. Raman quietly opened the bedroom door, pulled activation pins out of two canisters of a special knockout gas, and rolled them into the family room. The canisters spewed an invisible gas for a few seconds until an agent saw them and yelled loudly. The gas overpowered the safe house personnel in the family room, and despite their best effort at covering their faces and trying to run to safety, they all fell silently to the floor. One agent fell against the bathroom door blocking Allen's exit.

Allen's service in the army reserves included nerve gas training and he reacted instinctively at the first whiff of gas. He soaked a towel in water and stuffed it under the door. He soaked another and held it against his nose. How could he get to Ann? He unlocked the bathroom window and struggled to lift it without making any noise. It resisted at first, then slid up easily. He wriggled through it and fell to the ground in the dark. He soon discovered one of the open windows and crawled back into the house.

The strike team had seemingly overpowered the entire security force without firing a shot, which further pumped-up Raman's already bloated ego. Raman and Rambaud quickly teamed up with Assad and Attar to check the fallen victims as they searched for Kim Sang. Allen peeked out of the bedroom doorway, watching them, and wondered how he could overcome such a well-armed group of mercenaries. He could overpower one, maybe two, before the others fought back, and then he would be putting everyone at risk, especially Ann, who had passed out in the chair where she had fallen asleep and dreamed.

Harrows and three agents in the living room had not been surprised or knocked out. After the incident with the car backfiring in front of the safe house, he had ordered the agents in the living room to take out their gas masks as a precaution. They quickly took up defensive positions and were alert to any potential problem. When the agent in the family room saw the gas cylinders and yelled, the agents in the living room had instinctively put their masks on. Unknown to Harrows, or Raman's assault team, the air conditioner that was added to the garage when it was converted to living space was isolated from the rest of the house's air conditioning equipment. Kim Sang and the two security agents guarding him were also unaware of the gas in the house.

Assad rolled the fallen agent away from the bathroom door and pointed his mini-machine gun inside as he flung it open. He ignored the open window as he checked the shower. «Sang is not here, either.»

Raman motioned him toward the living room. "Check the front rooms." As Assad proceeded cautiously down the hallway toward the living room, Raman motioned to Rambaud. "Check the garage."

Rambaud jiggled the metal garage door. "It's locked!"

"Break it down."

Rambaud kicked hard on the door, but it wouldn't budge despite repeated attempts. In frustration he shot the lock off with his automatic rifle, kicked the door in and faced two agents pointing their automatic rifles at him. Both agents fired before he could squeeze off a single round. Rambaud was struck several times and fell backward against Attar knocking him down. Raman ducked down behind the chair where Ann was lying unconscious and began spraying a barrage of bullets into the open garage doorway.

Allen's heart almost jumped out of his throat at the sight of the stream of bullets flying right over Ann. Hot, empty bullet casings were cascading on her as he crept up on Raman from behind, pulled his gas mask off and choked him with an arm lock until he lost consciousness. Attar saw Allen pulling Raman's gas mask on as he struggled to his feet. He found his automatic rifle and pointed it at him. Allen pulled Raman in front as Attar fired a single shot at him. The bullet passed through Raman's chest and struck Allen in the shoulder knocking him back. Allen's grip pulled Raman on top of him as he fell backward.

Attar stood up and lurched forward to help Raman but was hit immediately by several bullets from the agents in the garage. Allen almost passed out from the pain in his shoulder but managed to pull Raman's rifle out of his death grip.

Assad approached the front living room cautiously. The lights were out, and he ran his hand along the hallway wall feeling for a light switch. He suddenly tripped over Tom Billings, Harrows second in command, who was kneeling and staring out a side window at the empty street in front of the house. Assad fell on top of him in the dark.

Billings jerked his gas mask off, pushed Assad away, and shouted at him. "Watch what the hell you're doing!" Assad scrambled to his knees. He searched frantically for his gun in the dark as the first shots rang out from the family room. Harrows flipped on the overhead light in the living room just as Assad grabbed his mini-machine gun. They saw each other at the same moment. Assad squeezed the trigger on his gun spraying the front room with bullets. Harrows and the other two agents in the room dove for cover. Billings shoved Assad from the back, and he fell to the floor again. Billings jumped on him, and they struggled for the gun. Billings rolled on top and jammed the gun against his throat to choke him. Assad started to lose consciousness. He summoned all his strength and slammed the butt of the gun into Billing's face. Billings fell on top of him in a daze. Assad pushed him off and rolled to the side as several bullets from Harrows ricocheted around him. He shot the overhead light and sprayed bullets randomly into the darkness. Bullets slammed into the sheetrock near his head powdering him with chalk dust as he struggled to his feet. He fired the last rounds in his machine gun as he backed down the hallway. As he entered the family room, he saw Raman lying on top of Allen. He heard a clicking noise, and realized his weapon was empty. He glanced up at the open garage door and saw the flash of an agent's gun. An excruciating pain filled his chest as he was knocked back into the hallway. Allen took a deep breath and passed out with Raman lying on top of him.

A strong smell of antiseptics greeted Allen as he woke up. He was lying on his right side on a bed. He felt a dull ache in his shoulder and grimaced. When Ann saw him open his eyes, she leaned over and kissed him.

"How do you feel?"

"I've felt better," he replied weakly.

"One of the agents knows first aid and managed to patch you up. He said the bullet passed through your shoulder without hitting any arteries." She saw Allen grimace from the pain. "He said to rest as much as possible, Allen."

"I'm afraid we won't have too much time for resting." Harrows was standing behind Allen, examining his wound. "You'll be okay in a little while. We'll need your services ASAP, Dr. Atkins."

Allen nodded wearily. "How are you?" he asked Ann, remembering her encounter with the knockout gas and the spent bullet casings falling on her.

"I still have a headache, and a few blisters for some reason, but I'm all right."

Harrows had walked around the bed and was standing with his arms folded as Allen looked up at him. "How many men did you lose?"

"We were lucky, one dead, and two wounded that should recover."

"What about the defector?"

Harrows shook his head in disbelief. "Not a scratch."

"What about the attackers?"

"The leader is still alive but in critical condition. The rest died. Several ambulances are on the way here."

"What do we do now?"

"Well, we have to clear out of here. They obviously know about this place and could try another attack. I know a place they won't think of."

Neighbors had poured into the street at the sound of the major gun battle that had raged briefly in the safe house. They stared with fascination as two helicopters landed in the street. A dozen people with a few pieces of luggage limped, ran, or were helped from the house to the waiting helicopters. The helicopters quickly lifted off into the night as a score of police, fire, and ambulances converged on the former "safe house."

The helicopters flew north toward the ancient city of Chiang Mai to a remote rendezvous point near a resort high in the mountains. The passengers quickly transferred everything to several black limousines waiting to transport them the remaining distance.

The five-star resort was almost surrounded by forest and was composed of a dozen buildings each containing six luxury suites. At a little after three in the morning, the convoy pulled quietly into the parking lot of a building located in a secluded area of the resort. Two members of the task force electronically swept the area and the team soon occupied one entire building. Harrows personally carried two of the four aluminum suitcases containing the lie-detector equipment into the new "safe house."

Ann had taken several tranquilizers and sat down to rest. Allen returned from the new "command center" kitchen area and handed her a candy bar and cold soft drink. She stared at the candy bar for a moment and suddenly stood up and put her arms around him tightly. He was a little surprised but hugged her back just as tightly. She put her head on his chest and he found himself caressing her hair.

"What is it, Ann?"

"I have a bad feeling about all of this... as if we won't get out of this alive. I know it's silly, but the feeling won't go away."

"I wish I could tell you that we'll be back in Houston soon, but I can't."

Something that had been nagging at the back of her subconscious suddenly became clear to her. Something Allen had said about missed opportunities. "Allen?" she said, snuggling into his arms.

"What?"

"I want to get married, right now."

Allen held her back to look into her eyes. "What?"

"If something happens to you or me, I want it to be as your wife."

Allen swallowed hard. He had never seen Ann like this.

"Ann, you're just feeling a lot of stress... and rightly so. We'll get out of this, I'm certain."

What's he afraid of? "I don't care. Will you do it or not?"

"Thailand's almost completely Buddhist and Moslem; it will be extremely difficult to find a Christian clergyman, and...." Allen's voice trailed off as he studied her face. She's serious.

From past experiences with Ann, he knew she rarely changed her mind once she made it up. This didn't seem like the right place or time, but he didn't want to upset her even more than she already was. And, somehow, the idea excited him.

"All right. I'll ask Harrow to see if it's possible."

"Thanks, Allen." She kissed him softly and he held her for a moment before he left to find Harrows.

When Hak Suhendra learned of the failed attempt to recover Sang and the demise of the assault team, he contemplated suicide. He had no more money to hire professional assassins. He assembled the small group of intelligence agents in his section and asked for volunteers to follow him into battle if necessary to eliminate Sang once and for all. Despite his exhortations and threats, he was only able to obtain a single volunteer.

CHAPTER 32

One of the last things Harrows had expected was a request for a marriage, but if it would help ensure their cooperation, what did he care? His aides used their considerable contacts and quickly found a clergyman willing to travel by helicopter and arranged a simple ceremony. The clergyman had anticipated a possible problem with wedding rings and brought along several gold bands. Allen and Ann found their ring sizes and Harrows stood in for Ann's father. Concern over the possibility of another attack, however unlikely, hastened the proceedings. Ann and Allen sealed their new relationship with a kiss.

Based on many years of experience, Harrows concluded they should utilize the «invisible» nature of Allen's system as a way of getting Kim Sang off guard and more susceptible to precise questioning. While Allen recuperated, Ann supervised several technicians as they worked the rest of the day to mount the spectrophotometer, laser, unidirectional microphone, and infrared camera into a bamboo cabinet they found in one of the suites. The equipment was so well hidden, Allen couldn›t see it when they showed it to him. Harrows was so pleased; he gave the team a break for the rest of the night.

Allen and Ann moved into an exquisite suite next to the new "command center." The room had a large balcony with a magnificent view of the mountains. Allen and Ann took advantage of every opportunity to be alone together.

It could have been the stressful conditions they were under or the need to find comfort in each other, but they soon settled into a mature and intimate relationship.

Despite a powerful painkiller, Allen woke up during the night with a sharp pain that almost seemed like someone was twisting a knife in his shoulder. He picked up his watch from a nearby nightstand and yawned. «Two-thirty!»

He sat up slowly in bed wondering why the light was on. Ann had fallen asleep reading a novel and was still holding it in front of her. Allen gently pulled the novel from her grasp and almost laughed out loud when he examined it. He never would have guessed that she would read romance novels. Ann was a forceful businesswoman with a determined style that impressed even hard-core executives that tended to downplay the role of women in business. Somehow, he wouldn't have guessed her reading preference.

I guess there's a lot I don't know about her. He also wouldn't have expected his inadvertent remark about her being pretty would rekindle an old flame he had long assumed would never be re-ignited. Even more unexpected was the virtual kidnapping by an intelligence agency that had placed them in great danger.

The most surprising thing of all was Ann's sudden request they get married. He shook his head as he watched Ann sleep. He smiled when he saw her wearing one of his NASA T-shirts. Harrows had sent two agents to the hotel in Bangkok to retrieve their luggage, and Ann had several nightgowns with her. It seemed a little romantic that she preferred to sleep in one of his shirts. He leaned over and kissed her, turned off the light and snuggled next to her.

Ann woke up sweating, wondering why she was so hot. A ceiling fan turned lazily overhead, and a cool gentle breeze was blowing in though a screen in the open balcony door. Vertical blinds near the balcony door were swaying back and forth, and a steady cacophony of bird and small animal night sounds seemed soothing.

Allen had wrapped his arms around her and intertwined his legs with hers. It was endearing but too warm for comfort. She was also slightly allergic to an aftershave Allen used that seemed to endure long after he shaved. She managed to untangle herself from Allen and sat up. The shirt she had borrowed from him was soaked in sweat and she slipped out of bed and quickly changed into another of Allen's shirts. After a quick trip to the bathroom and a large glass of water, she laid down next to Allen listening to his rhythmical breathing.

Like a majority of men his age, Allen was sleeping in his briefs. That fact brought back a memory of Phil buried deep in her subconscious and she smiled.

"I wonder if it would work on you," she whispered. She leaned over Allen and began licking his nose and lips. After a few minutes, she chuckled softly. "Dear Abby's readers should know about this."

Kim Sang had also settled into a comfortable routine. Although he was kept under continuous guard, he was allowed some freedom to roam about. Harrows was waiting for an opportune moment to question him with the voice-stress portion of the lie-detector system in operation. The bamboo cabinet had been carefully located in his bedroom along with a comfortable chair overlooking the balcony. Sang had thought it a little odd the chair had been nailed to the floor, but often sat in it to relax. Harrows experimented with an indirect approach at first by locating a telephone on a small table next to the chair and having his chief interrogator call Sang periodically to ask a few additional questions 'for clarification purposes'.

The lie-detector system provided a hint that Sang may not be completely telling the truth. Many of his answers contained elements of truth and lies and Harrows was becoming frustrated. His biggest concern was the possibility the North Koreans would locate his latest hideaway before they could determine the truthfulness of Sang's story. The resort suites would be extremely difficult to defend.

CHAPTER 33

Something seemed wrong to Allen. He couldn›t quite place it, something in Sang's answers, maybe. He was about to discuss his gut feelings with Ann as Harrows paid an unexpected early morning visit to Allen and Ann's room. The sun was just starting to rise as Allen answered Harrows› insistent knocking.

"Yes, Mr. Harrows?"

"May I come in?"

"Of course."

Ann had been in the bathroom and came out, just as Harrows entered.

"What is it? What's going on?" she asked, concerned.

"I'm afraid we don't have very much time. We have intelligence that suggests the North Koreans have discovered our location and are on the way here."

The wound in Allen's shoulder seemed to start throbbing at the thought of another encounter. "How much time do we have?"

"Our best guess is noon tomorrow. We could be out of here in minutes if we had to, but we can't keep running away forever. We need a definitive answer on Sang ASAP."

"What do you want us to do?"

"Before I answer that, I have a surprise for you." Harrows turned and opened the door. Phil had been leaning against the hallway wall opposite the door and entered quickly when Harrows motioned to him.

"Phil!" cried Allen and Ann together. Ann rushed to hug him. Allen waited a moment and shook his hand. "I'm sorry you had to get involved, but I'm sure glad you're here."

"Do you know what's going on?"

Phil nodded. "Mr. Harrows filled me in."

"We managed to find him just in time," chuckled Harrows.

"Where were you, Phil?"

Phil glanced at Harrows and laughed as he replied. "Joanne and I were on Pitcairn Island."

Ann frowned. "I know I should know that name, but it just slips my mind."

"Mutiny on the Bounty!" exclaimed Allen.

"Joanne is a big fan of the book and movies. She always wanted to go there. I told her she could pick the place, but I didn't guess that one."

Adam wondered how they located Phil and Joanne on such a remote island. "How did you find them, Mr. Harrows?"

"We tried all the usual stuff, checking all the airline reservations out of Los Angeles, along with car rentals, interstate bus lines, AMTRAK, even the ferry to Catalina Island, with no luck. We even checked for reservations under Mrs. Conley's maiden name with no luck. We didn't think of looking for Mr. and Mrs. Christian."

Phil chuckled as Harrows sat down with a smile on his face.

"We finally put out a news story that they had just won a big lottery and the news media was trying to find them. It said there was a large reward for information leading to them. Someone on the island had a satellite receiver and called the number on the screen immediately."

"That's a new twist on intelligence gathering," commented Allen.

Ann surveyed Phil in the early morning light. "You look just fine. How's Joanne?"

"Joanne's fine. She said to tell both of you 'hi'."

"Where is she?"

Harrows answered for Phil." She's safe with your children, Mrs. Atkins. It seemed safer that way, and they can take care of her as well."

Phil couldn't have been more surprised. "Mrs. Atkins? When did you two get married?"

Ann blushed. "Two days ago."

Allen smiled at the shocked expression on Phil's face. "We'll tell you all about it when things aren't so hectic."

Allen closed the door and stood by Harrows. "Well, now that we're all together, Mr. harrows, what's next?"

"As I explained, we are under a tight schedule to determine whether Mr. Sang is for real or a double agent."

Phil's expression reflected his confusion. "Of course. But what can I do? I don't speak Korean.

"We have translators. Do the same thing you did at the trial with those witnesses that were lying. We've discussed the information we need."

Phil walked to the open balcony door and paused to enjoy a cool morning breeze before he replied. "Why would this man say anything to me, he wouldn't say to any other interrogator?"

Harrows replied softly. "We will tell him you are an Assistant Secretary of State and have the power to grant him asylum or turn him back over to the North Koreans."

Phil spun around in surprise. "What? Why me? Why don't you use one of your own men?"

"We've been studying him carefully. He seems to know when he is talking to someone in the military or an intelligence officer. It's almost like he has a sixth sense. He would undoubtedly be more at ease with you. He wouldn't know anything about the lie-detector system since it's pretty much invisible now. He tenses up every time he sees anything that looks like a lie detector. And...you also know more about questioning a subject based on its output than anyone, even Dr. Atkins."

Allen chuckled. "He's got you there, Phil."

Ann was standing next to Allen. She put her hands around his arm and moved closer to him. "It's pretty early, Mr. Harrows. What do you have in mind?"

"It may sound odd, but I'm going to tell him the truth. We think the North Koreans know he is here and are on their way. I'm also going to tell him we need to ask him one final set of questions, and by God, he better tell us the truth!"

They all laughed.

"We probably don't need you right now, Mrs. Atkins. You can wait here. We'll send for you if we need you. Gentlemen, he's waiting in his room. Shall we go?" Harrows walked to the door and held it open for them. Allan kissed Ann. He exchanged a knowing glance with Phil, and they followed Harrows toward Sang's suite.

CHAPTER 34

Kim Sang was pacing nervously in his bedroom. He could sense something was about to happen from the sudden increase in activity in the agents guarding the building. He watched in amazement as two black sedans arrived and quietly disgorged ten additional agents. Maybe his former colleagues knew where he was. They had almost succeeded in killing him in their last attack at the safe house. His level of regret at his decision to seek asylum had reached a new high when there was a knock on the entry door. He quickly opened the door and found Harrows, Michael Gore, Roger Douglas, his interpreter, and a new face. Harrows› interpreter informed him they needed to speak to him immediately. Sang opened the door further, bowed to Harrows, and backed up for them to enter.

The interpreter informed him that Phil was an Assistant Secretary of State and he had the power to grant him asylum or return him to the North Koreans. Sang breathed a sigh of relief.

"At last," he said to himself.

"Let's go into the bedroom, it's more comfortable there," suggested Phil through the interpreter. Phil motioned Sang to sit in the chair as he and the interpreter sat down on the bed. The only chair in the room had been firmly anchored to the floor to ensure alignment of Sang's face with the laser and infrared camera. Harrows, Gore, Douglas and two security agents waited in the living room, listening to every word.

Phil began by asking him some routine questions to validate the system's operation against known facts.

In the next suite, Allen had been watching the VSA analysis, and the infrared camera window, when Sherlock attempted to contact Allen.

"Allen?" appeared in large letters. He turned on the speakers. "Yes?"

"Allen, I don't understand some of the questions being asked, or the immediate response of the subject."

Allen glanced at the voice-recognition window, and indeed, the Korean portion of the interview was filled with question marks and an occasional guess at an English word by the voice-recognition software.

"The questions are being translated from English into Korean and the subject is answering in Korean. An interpreter is translating the subject's answer into English."

Sherlock didn't respond for a moment. Allen watched the screen as Sherlock broke his answer into several blocks of words, reviewed the dictionary definition of several words, and pieced the answer back together.

"The voice-recognition software has modules for French, Spanish, Russian, and German, but not Korean. What do you want me to do?"

"Ignore the translated question, and try to analyze the answer in Korean, but record the interpreter's answer."

Sherlock broke Allen's answer into several pieces but couldn't seem to fit them back together. "I don't understand," replied Sherlock.

"Record the question asked in English and use the voice stress analysis, the infrared imaging and the spectroscopic analysis on the subject's answer. I know it's in Korean, but the principles of stress should still be valid. And of course, record the subject's answer after it has been translated into English."

Sherlock analyzed Allen's answer and after a few moments simply replied, "Understood."

Phil studied a list of questions that had been prepared for him by Harrow's men. He glanced up at Sang, who smiled at him. «And what was your position within the KRG?»

"I was responsible for the development of several weapon systems that had the capability of delivering nuclear warheads." Sang didn't blink as he watched the interpreter translate his answer.

Phil almost jumped when Sherlock's voice boomed into his earphone.

"The probability of truth was 20% for the last response of the subject, Sang."

Why was Sang clearly lying when he knew this was his last chance? Phil glanced at the next few questions on the list and tossed the clipboard onto a nightstand by the bed. Sang was intently watching every move Phil made.

"How old are you?"

Sang seemed surprised at the question. He frowned. "Thirty nine."

"It's been my experience that someone would have to work a great number of years to be in charge of such an important development program. I don't think you're old enough."

Sang again seemed surprised at Phil's questions and statements that seemed directed at him instead of the information he wanted to trade for asylum. "I tell you; I WAS the director of nuclear weapons development."

Sherlock reconfirmed the lie and Phil stood up, paced back and forth a few times and stopped in front of Sang with his arms folded. "No, you weren't."

Phil looked for a change in expression, hoping Sang would give some indication he had been caught in a lie, but his face held no expression. He didn't reply and Phil continued.

"I don't believe you, and neither do those men," he said, gesturing toward the living room where Harrows, Michael Gore and Roger Douglas waited.

Phil walked to the bedroom door and motioned Harrows to come into the bedroom and spoke softly to him. Harrows removed his handgun from a shoulder holster and handed it to him. While still facing Harrows, Phil removed the ammunition clip and emptied it. He pushed the clip in until it almost locked into position, turned and tossed the handgun to Sang who caught it in midair.

The clip had almost fallen out and Sang automatically shoved it back in until it locked in place. He glanced at the gun and tossed it back to Phil. He frowned when he saw Phil smile at him. He suddenly realized Phil was testing him. It took every bit of his will power, but he managed to betray no emotions and sat unmoving in his chair.

Phil tried not to react when Allen spoke through his earphone.

"Phil, this is just a guess on my part, but I think he understands English. He reacts immediately to your questioning, even before he hears it translated. This is just a guess on my part but keep it in mind when you are questioning him. Watch his response when you ask the question."

Harrows stood by the bedroom door with his arms folded in frustration as Phil paced around the room. "Are you married?"

Sang was getting confused. Why would they want to know these things? He asked Phil why. Phil was still trying to establish a base case of true answers for the lie detector, but he couldn't tell Sang that.

"We were just wondering if they could hold your family as a bargaining tool to get you back," he lied.

That was a weak but believable answer. Sang finally looked up at Phil. "I would like to speak to you alone." Harrows and the others were not pleased when Sang's answer was translated. Phil casually handed Harrows his gun back and whispered in his ear.

"Let me see what he wants. You can follow what's going on with Allen."

Harrows was visibly upset, but he decided to follow Phil's advice. He had spent quite a bit of money and had called in some old debts with various other agencies to find Phil. He would have to trust Phil's instinct in this. He motioned to Gore, Douglas, and the security agents, and they left the suite.

Phil remembered Allen's deduction on Sang and motioned for the translator to leave as well. He seemed confused but left quickly. Sang waited patiently for Phil to shoo the remaining guards out of the suite and close the entry door. Phil stood in front of Sang with his hands in his pockets.

"Why are you really asking for asylum?"

Sang answered in English with almost no accent. "I don't know who you really are, but I think you are not CIA, or even a part of the intelligence community. I am also certain you are not with the State Department. I have to know who you are before I can say anything else."

Allen had been right about Sang, and Phil didn't react to Sang's command of English.

"I'm just a lawyer who was on his honeymoon, until you decided to defect."

Sang's expression didn't change. "No, I think there is more. The CIA has many people who could ask the questions on that list."

Phil knew Harrows and the others were listening but Sang didn't.

"You're right. A friend of mine invented a new lie-detector system, and we used it at a murder trial to solve the case and find the true murderer. They wanted to use it to see if you are telling the truth."

Sang studied Phil before answering. "I have seen the equipment. It's clever, but not foolproof."

Phil was a little relieved that Sang thought the system hammered together by Harrow's men was the only one he knew about. But who was he really?

"You know too much about weapons, lie detectors, and intelligence to have headed a nuclear weapons development."

Sang stood up and motioned Phil to follow him to the balcony. Phil knew he was on his own. The lie detector couldn't pick up a quiet conversation on the balcony over the early morning sound of birds and a nearby decorative fountain.

Allen spoke urgently in his earphone. "Phil, don't go out there with him. It could be dangerous, and we can't hear what's going on from here." Phil ignored him and they stood looking out at an early morning fog that shrouded the nearby forest and mountains without speaking for a little while. Sang moved closer to Phil and spoke softly.

"You are right. I know almost nothing about weapons development. But I do know a great deal about the Korean intelligence system, and the fact that my room has listening devices."

He caught Phil off guard, but Phil gave no outward sign.

"You are probably wondering why I am telling you this instead of the CIA."

"That was one of several questions that came to mind."

"I really did intend to defect, but when I met my counterpoint to begin discussing what I could offer in exchange, I recognized him by his voice as one of our contacts within the CIA. Fortunately, he didn't recognize me."

Phil felt a cold chill run down his spine as Sang's latest revelation sank in. That meant one of the three men with Allen was a double agent.

Phil leaned closer to him and whispered softly. "Who is it?"

"I will tell you that if you can tell me how you will keep him from murdering me."

They stared at each other for a minute.

"You'll have to trust me on this," replied Phil thoughtfully.

«I don›t like this,» growled Harrows. «I›m going in there.»

"Wait," exclaimed Allen. "You have to give him a chance."

"I don't have to do anything." Harrows had his hand on the butt of his pistol.

Allen corrected himself. "I know you're in charge here, Mr. Harrows, but you've invested a lot of time and money to find out the truth. Why don't you see what Phil can learn from him?"

"I agree," said Gore. "He's here for a reason, and the North Koreans sure want to get rid of him. Why don't we see what Mr. Conley can do?"

Harrows took a deep breath to calm himself down. "All right. I'll give him five more minutes, but that's all." Harrows pulled out a radio and doubled the guard on the grounds near Sang's suite and balcony to prevent a possible escape.

Allen realized the system was picking up a noise that sounded like tapping. Phil was signaling to him in Morse Code. He strained to pick out the sequences, but it didn't make sense at first, then it slowly dawned on him. Phil was communicating in Esperanto. They had both been in the Boy Scouts, and that was the language they had used at all international meetings. He closed his eyes to concentrate.

"What in the hell is that noise?" Harrows bellowed. Phil and Sang were still out of sight of the cameras.

The answer suddenly came to Douglas. "Morse Code!"

Gore picked up a notepad and tried to write the message, but it didn't make sense. "I don't get it. It's not in English." He showed the letter sequence to the interpreter who shook his head. "It's not Korean either."

Allen felt his heart beating a little faster when the message from Phil ended.

"SANG IS IN INTELLIGENCE. ONE OF THE MEN WITH YOU IS A DOUBLE AGENT. PROTECT OTHERS WHEN HE REVEALS NAME."

The others were all gathered around the laptop speakers trying to hear what Sang might be saying on the balcony. Allen backed up slowly and they moved in even closer to listen.

Sang walked back into the room with Phil and stood near the chair. Phil nodded to him.

"So you see, Mr. Conley, I could not tell anyone that Mr. Douglas is an agent of ours for fear that he would kill me."

Phil motioned him to go back to the balcony.

Everyone gathered around the laptop froze. Douglas happened to be standing next to Harrows. In a single motion, he grabbed Harrows' gun and stepped back from them pointing the gun at Harrows, Gore, and the interpreter.

"DAMMIT! Don't move or I'll shoot."

"You son of a bitch. Give it up, Douglas!" demanded Harrows angrily. "You can't possible get out of here, even if you killed us."

"You're wrong, JOSEPH! No one outside of this room knows anything. I could walk right past them before they found your bodies."

"You lousy traitor!"

"You wouldn't believe the truth if you heard it. Everyone has their price. Yours is just a little higher than mine, that's all."

Douglas kept glancing at Gore and the interpreter, and Harrows calculated the odds of getting a shot off before Douglas but found the odds unacceptable. He needed a distraction. "What do you want?"

"I want you to provide me with safe escort out of here. But first, I'm going to take care of that little stool pigeon."

Allen hit Douglas with a hard karate chop at a precise location on the back of his neck. Douglas fell like a rock and Allen kicked the gun away. They all breathed a little easier.

"We owe you a great deal, Dr. Atkins."

"YOU CAN PAY ME OFF BY GETTING US THE HELL OUT OF HERE!"

Hak Suhendra's source of inside information had proven extremely accurate, and he trusted it without hesitation. He utilized his contacts in Bangkok to recruit a large number of gang members to help in one last desperate attempt to kill Sang. Although he had no money to offer them, they eagerly volunteered when promised drugs and guns in exchange for their help.

CHAPTER 35

His mission accomplished, Harrows ordered a quick evacuation of the resort. He didn›t have to worry about drawing attention to them now. Four helicopters landed in the parking lot and were loaded in record time. When they were airborne, Harrows keyed the send button on his radio.

"Carb One to Carb Two!" he yelled over the drone of the helicopter.

Pilot Ben Smith took a deep breath. "Carb Two here, sir."

"Let me speak to Dr. Atkins."

"Who, sir?"

"Dr. Atkins, one of the consultants helping us!"

"Uh... we don't have any consultants on board, sir. Only Carborundum personnel."

"What! Say that again!"

"We didn't receive any instructions to transport the consultants, sir!"

"Carb Two, return to the resort immediately and bring everyone you left behind and their luggage. That's an order! If anything happens to them, I'll hold you personally responsible."

"Yes, sir," replied a nervous Ben Smith.

The fog was beginning to lift off the mountains as Allen and Ann finished packing their suitcases.

"Allen, what about the lie-detector system?"

"They can have the damn thing. I just want to get back to the U.S. and start a new life with you. He closed his suitcase and set it by the door.

Ann stared at him for a moment then put her arms around him and kissed him passionately. As Allen began to hug her there was a knock on the door. Allen let go of her and opened the door for Phil.

"Ready to go, guys?"

Ann placed her suitcase by the door next to Allen's. "We've been ready since we got here."

"I saw Harrows a little while ago. He said the helicopters will be here any minute. It's strange though, I thought I heard some helicopters come and go a few minutes ago." He shrugged his shoulders. "Oh, well, do you need help with your luggage?"

Ann picked up her purse "They said someone would come for them. I hope they didn't forget."

Ann grabbed Allen's arm when an explosion in the distance rattled the windows. "What was that?"

Allen pulled her out the door. "Come on, let's go now!" They ran outside and found the resort deserted. A limousine was parked nearby and Allen ran to it, opened the door, and glanced inside.

"THE KEY'S ARE IN THE IGNITION, COME ON!"

"What about our luggage?"

"It's too late for that, Ann, come on!"

As Phil and Ann ran for the limousine, there was another loud explosion in the distance followed by a rumbling noise. Allen started to open the door for them but stopped. The noise quickly got louder until Allen yelled. "Damn. We're too late."

Ann looked at him questioningly. "What?"

Allen grabbed her hand and pulled. "Quick, back to the resort."

"Not another attack?" Desperation was creeping into her voice.

He nodded. "Come on."

They ran for the building containing the suites they had occupied and were thrown to the ground as a barrage of mortar shells slammed into the field around them, one hitting the limousine. An enormous ball of flames rose into the sky as they ran into the house.

"What now?"

Allen glanced out the window. "We need some weapons."

Phil remembered the car. "Maybe there's some in the trunk of the limo."

"It's too dangerous to run out there now. Even if we found them, they'd probably be messed up and too hot to use."

"Do you think this place has a basement?" Phil was looking around the building for a door he hoped would lead to a basement."

"I doubt it."

"Where can we go, then?"

"This may sound stupid, but I think we would have a better chance in the forest."

"What!" Ann and Phil replied together.

"All the mercenaries will be coming to this building. We might have a better chance in the forest. We might only run into a few that way."

"If we do, how will we defend ourselves? We don't have any guns," said Phil, nervously.

"We'll have to make do. Come on."

Hak Suhendra personally directed the attack, which was his last chance to kill Sang. Capture wasn›t even an issue. He just hoped Sang had been trying to make a deal and not divulged very much of his considerable knowledge of field agents located all over Southeast Asia.

He was amazed at the arms in possession of the recruited gang members, some it was newer and better than the armaments available to the North Korean military! He felt confident they could finally take out Sang and as many of the intelligence agents that were guarding him as possible.

Special agent Harrows had been in contact with the transport helicopter sent to retrieve Allen, Ann, and Phil when the pilot reported seeing mortar shells hitting the resort and the limousine parked outside. The helicopter came under fire from the ground and the pilot backed the helicopter away from the resort and waited for further orders. Harrows swallowed hard and ordered the helicopter with Sang and Douglas to proceed to a designated rendezvous point. The rest were ordered to go back and try and effect a rescue.

Phil and Ann followed Allen to the back of the building. They bent over and ran for the forest that was less than a hundred feet away. Allen found a dense patch of underbrush and he motioned them into it. As they tried to cover themselves with leaves and vines, they heard yelling and gunfire from automatic rifles.

Several tough-looking gang members ran past them. More than three dozen gang members advanced toward the building from all directions and quickly occupied it.

Allen peeked through the underbrush and waved to them to follow him. They walked as quickly and quietly as they could through the dense undergrowth when Allen suddenly stopped. They stopped behind him and saw a mercenary squatting on the ground with his back to them. Allen smiled at them and motioned them to stay where they were. They watched him creep up on the mercenary and hit him on the head. Allen returned quickly with an AK-47 automatic rifle and several ammunition clips.

"It's not much, but it's a start."

"Which way?"

"Away from the house, that's for sure."

Allen began to move quickly through the dense underbrush, with Ann and Phil following closely behind. Allen suddenly stopped and motioned them to get down. They saw Allen put the rifle down and move quickly forward.

Two oriental intelligence agents dressed in camouflage clothing were standing next to an off-road vehicle that looked a great deal like a jeep. One was speaking angrily into a portable radio. Allen found a branch of a tree and waited for them to look away. He moved quickly, covering the last few feet in the air as he executed a flying kick and knocked both of them over. Hak Suhendra recovered quickly and tried to aim his rifle at Allen, but Allen knocked it away with the branch. He quickly kicked Suhendra in the face and whirled around kicking the other one who was struggling to get up. Allen hit him with the branch, breaking it over his head. He searched the fallen agents for keys for the jeep but found none.

"Damn!" he said to himself.

Phil had been watching and ran over to Allen with Ann following closely.

"They don't have any keys," Allen said softly to them.

"That's not a problem."

Phil opened the door of the vehicle and reached under the steering column tugging at something. Allen and Ann had followed him and were watching him when he looked up.

"Do either of you have a knife?"

Allen walked over and began searching the unconscious agents. He brought back a large knife with a wicked looking blade.

"How about this."

Phil chuckled, took the knife, and cut some wires. In a moment, he touched two wires together and the vehicle coughed into life.

Ann watched in amazement. "Where did you learn to do that?"

"I once prosecuted a gang member for auto theft. He claimed you couldn't hot-wire a car in the time frame we said he did it. So, I searched the prison records and found an inmate who gave a demonstration in the parking lot of the courthouse. He did it in less time than we were accusing the gang member of. Needless to say, we won the case."

"That's really interesting. But we need to get going," urged Allen.

They spotted a path through the forest and pushed through the brush as fast as the four-wheel-drive vehicle would let them.

The three agency helicopters attacked the resort building with machine guns and rockets, wiping out most of the gang members before the rest fled. Two helicopters landed and two teams of agents scoured the resort for Allen, Ann and Phil.

"Carb Two to Carb One!"

"Carb One, here."

"They're not in the house, sir. And we can't find any horizontals resembling them."

"Where in the hell did they go?" Harrows began to feel a sense of desperation when the other patrol leader reported in.

"Carb Three to Carb One."

"Carb One, here."

"Sir, we found an unconscious horizontal several hundred yards from the compound and he's clean. The consultants may have surprised him and taken his valuables."

Harrows chuckled to himself. Atkins! "What are the coordinates?"

"They would seem to be heading north from the compound, sir."

Harrows ordered the search teams back to the helicopters and they spread out in a search pattern heading north from the compound.

A helicopter buzzed the off-road vehicle as it bounced through an old, overgrown forest trail. Allen stopped the vehicle.

"What is it, Allen?"

"Helicopters. They may be ours, but I'm not sure. Maybe we should wait here a while, just in case."

"What if the mercenaries find us?"

Allen considered the risks and consequences. "Let's go for it."

He shoved the stick into low gear and the vehicle lurched ahead, dragging vines and underbrush. Allen had a hard time seeing ahead as overhanging vines and tree branches slammed into the windshield. They suddenly veered into a large clearing. Allen jammed on the brakes as a helicopter bristling with machine guns dropped down in front of them. He sat back with relief as he recognized a large Carborundum logo on it.

Several agents jumped out and ran toward them. One jerked Allen's door open.

"Dr. Atkins, Mrs. Atkins, Mr. Conley. You must get on board immediately. There's a large number of gang members roaming the area."

They didn't have to be told twice. Agents helped them get on board and the helicopter rose and raced to join the others as they headed for a new rendezvous point.

The helicopter settled next to the others in what appeared to be a small, abandoned farming village. An agent led Allen and the others into a thatched roof hut where they saw Harrows talking on a portable radio unit. Harrows saw Allen but didn't have a chance of blocking Allen's full-frontal assault. Harrows hit the floor and was knocked unconscious. Several agents tried to grab Allen, but he kicked, flipped, and tossed them like dolls blocking all their efforts to constrain him.

Ann screamed to him. "Allen, don't! Are you out of your mind? He saved us."

Allen ignored her and had Harrows neck in a death grip as he opened his eyes.

It was a struggle for Harrows to breath, let alone talk. "Are you going to kill me, Doctor?"

"I should, you son of a bitch!"

Allen felt a gun press into the back of his head. "Let him go, or I'll pull this trigger, Atkins." Allen recognized the voice of Tom Billings, team leader of the security agents. He let Harrows go and stood up.

After rubbing the back of his head, and taking a deep breath, Harrows stood up. His men had Allen's hands tied behind his back and two were pointing handguns at his head. Ann had her hands over her face crying from the tension of the day's events and Allen's current situation. An agent was even holding a weapon on Phil as he hugged Ann to comfort her. Allen glared at Harrows. The other men Allen had knocked down were slowly getting to their feet. Harrows motioned them away.

"You must be thinking I left you behind on purpose."

"The idea had crossed my mind."

"Why would I do that?"

"We know too much. If the mercenaries killed us, you'd be rid of one future security problem."

Harrows shook his head. "The three of you were supposed to have left first, but the order got fouled up somehow. We were already in the air when I found out you had been accidentally left behind. I immediately ordered the crew to go back and get you. I was even talking to the pilot when the mortar shells hit the resort and the limousine. I can assure you, Dr. Atkins, I had no intention of leaving anyone behind."

The factual tone of Harrows' answer seemed to drain the anger out of Allen, and he slumped back against the wall behind him. Harrows motioned his men to lower their guns and untie his hands. Ann ran to hug him. When she stopped hugging him, Phil put his hand on Allen's shoulder in sympathy.

Harrows rubbed the back of his head, picked up the microphone, and resumed a report to his superiors on the success of the mission.

Allen looked up sheepishly at Ann and Phil. "I guess I looked like an idiot... again."

Phil laughed. "Don't be ridiculous. You saved our lives."

"We wouldn't have made it out of there if it weren't for you, Allen," Ann said softly. She kissed him on the forehead.

Phil tapped him on the shoulder. "What now?"

Allen glanced at Harrows. "I honestly don't know."

Another short helicopter ride brought them to the former rendezvous point, where several ordinary taxis met them to take them back to the same hotel they had occupied when they first arrived in Bangkok. After a few hours of resting and cleanup, they were whisked back to the international airport where a jet very much like the one they boarded in Los Angeles was waiting for them. The ramp was lowered, and Harrows was waiting for them by the stairs with Gore and Billings. Ann and Phil shook their hands as Allen walked awkwardly to Harrows who held out his hand. Allen gingerly shook it.

"I'm really sorry I attacked you like that, Mr. Harrows," he began.

"Forget it, doctor. I owe you and your wife and Mr. Conley quite a lot for helping me find the truth about Sang... and Douglas. I'll get some rather large 'ataboys' for this." He rubbed the back of his head. "But this sort of settles the score."

Allen laughed. "I guess you're right. Good luck with Sang. I hope he helps you find the rest of the double agents."

Harrows nodded and as Allen started to follow Ann up the ramp, he wished them luck on their next mission. Allen stopped to stare at him but was afraid to ask what he meant. The ramp was retracted, and the cabin door closed. The jet moved to the end of the runway and Harrows watched it take off and disappear in the clouds.

He turned to his team leaders. "Let's put Douglas on ice and get the hell out of here."

His men scurried to the waiting helicopters.

The co-pilot of the flight was a remarkably pretty Thai national. Allen and Phil couldn›t help gawking at her as she came out of the pilot's cabin to take their drink orders and to inquire if they wanted something to eat. Ann noticed and jabbed Allen. He cleared his throat.

"Do you know where we are going?"

"Japan!" she replied.

"Japan?" Ann's face registered her surprise. She put her hand on Allen's arm. "Then... Houston?"

"I certainly hope so." He decided not to mention Harrow's parting comment.

CHAPTER 36

Their jet landed at Narita airport, but it didn't taxi to the terminal. It proceeded to a remote building with a hanger. Ann was looking out the window and pointed to the large "Carborundum" sign over the door of the open hanger. The jet stopped and the engines wound down. A moment later, they felt it being towed into the hanger. When it stopped, the cabin door opened, and several agents entered. A youthful agent with blond hair and blue eyes entered to greet them.

"Good afternoon, Dr. Atkins, Mrs. Atkins, Mr. Conley. My name is John Bradley. We've been expecting you. How was your trip?

"As comfortable as a 56-inch cabin will allow," replied Phil, who had banged his head several times on the ceiling.

Bradley ignored the remark. "I have a car waiting to take you to a hotel where you can rest."

"When are we going home, John?" Ann was hopeful but not optimistic.

He seemed embarrassed that he couldn't answer the question. "I don't know, ma'am. My instructions were to bring you to the hotel."

There didn't seem to be any reason to talk to Bradley, so they scooted past him to the exit. Allen looked for weapons on the nearest maintenance personnel or security agents but couldn't spot any as they walked to a black sedan waiting for them.

The trip to the hotel in downtown Tokyo was uneventful. Ann, Allen, and Phil were bone tired, but the sights and sounds of Tokyo seemed to rejuvenate them. The sedan pulled into the drive of a luxurious hotel in downtown Tokyo, and they were met by several agents wearing suits. The impeccably dressed hotel clientele and staff stared at their soiled and wrinkled casual clothing as they walked through the lobby. They continued past the elevators to an escalator that led to a small but lavish conference room on the third floor. The agents escorting them closed the doors as they left.

«What now?» They settled into large, overstuffed chairs surrounding a marble conference table. Phil saw some cookies and coffee on a small table and proceeded to help himself. After a few minutes, Ann and Allen joined him.

A door opened, and three agents entered. One of them introduced himself and the others.

"Dr. Atkins? My name is Mark Dunlevy, and this is Jules Kroft and Tom Arens." He gestured toward the others. Allen bent over and rubbed his eyes as fatigue began to sap his energy. He was certain Ann and Phil were just as tired.

"I don't mean to be rude, Mr. Dunlevy, but we haven't had much to eat and haven't slept in the last two days. I hope you are here to tell us the number of the next commercial flight from Tokyo to Houston."

Dunlevy shook his head. "Sorry, I wish I could. You are here because we need your help on a rather difficult matter that involves national security."

Allen put his head in his hands. "Oh, no... here we go again."

Dunlevy continued. "I had the privilege of reading Joseph Harrows' report and was fascinated at the role your new lie-detector system played in uncovering the truth about Kim Sang. I'm sure he was just as appreciative."

Phil was too exhausted to care. "What do you want... Mark, isn't it?"

"I'm afraid I'm not at liberty to disclose that information." He looked at his watch. "However, our liaison officer, George Grant, can and he should be here at any moment."

Allen fidgeted in his chair. Liaison officer? Liaison to what... the government? That would make him a government bureaucrat, wouldn't it?

Grant arrived almost as if on cue. He was an imposing figure at six and one-half feet tall, with hazel eyes and blond hair cut so short it was almost invisible. He had a small group of support staff with him, but he was introduced to Allen, Phil, and Ann by Mark Dunlevy. Allen was as weary as he was wary of this huge, imposing figure, whom one might guess was used to having his way.

"Good day, it's a real pleasure to meet you. I was as fascinated by Joseph Harrows' report as I'm sure Mark Dunlevy told you he is."

Allen was too tired to be diplomatic to a government bureaucrat. "Mr. Grant, can we cut the bullshit and get to the point here? We haven't slept in two days. Why are we here, and what do you want?"

Grant didn't seem bothered by Allen's comments or questions. He motioned them to sit down at the conference table. He dismissed most of his staff and sat down next to Dunlevy. An assistant put a coffeepot and a plate of pastries on the table in front of them and left quietly.

"Considering all that the three of you have gone through, I guess that's the least I can do."

He paused as if putting a speech together in his mind.

"We have an urgent need for your new lie-detector system. In the broad scheme of things, it's much more important than the true identity of a defector. I know you may be too tired to even want to discuss this, but it is a matter of utmost urgency."

Allen yawned broadly at Grant. "You can have the system, Mr. Grant, we just want to go home."

"Yes... however, the three of you are the only ones that really know how the system works... its strengths and its weaknesses."

"We could train some of your agents...."

"There isn't time for that, unfortunately. Let me say one more thing about this. I can promise you this will be a very short effort on your part... probably less than two weeks. It won't be under such dangerous conditions either. In fact, compared to your last assignment, this will seem like a well-earned vacation at a luxury resort."

Allen, Ann and Phil just stared at him, too tired to think... too tired to object... even too tired to agree. After a few minutes, Grant leaned forward in his chair.

"Are you still awake?"

Ann rubbed her eyes and stifled a yawn. "Yes, Mr. Grant. I guess we were just wondering if there was any point of saying no."

Phil alternately studied the other agents' faces looking for a reaction, but there was none. "What if we did say no?"

"We really would prefer you volunteered for this assignment. I could make it worth your while, financially."

Ann was too tired for this incentive. "We don't need any money, Mr. Grant."

Grant seemed a little surprised. "I've never met anyone who didn't think they needed more money."

"We just made a deal with a major financial investor to exploit the lie-detector system. He is going to form several new companies and we are his partners."

Grant sat back in his chair. "Oh... yes. I heard about that. You don't seem to understand, Mrs. Atkins. Various intelligence agencies of the United States have seen the potential of this system and the benefits it could derive, especially in this area. We could easily block any proposed commercial venture based on it."

It didn't take very long for Grant's implied threat to sink in. Everything they had planned on, or assumed, could be flushed down the drain if Grant or his superiors followed through on his implied threat. He saw their reaction and held up his hands defensively.

"I didn't say that WOULD happen. I just said it COULD happen."

"There really isn't any difference, is there?" Ann replied sarcastically.

"Let me repeat something I said earlier. We really want you on board with us in a matter that is of the utmost importance, and is in the interest of the United States... of which you are no doubt loyal citizens."

Allen's temper flared up and he almost managed to control it, but not quite.

"How DARE you question our loyalty after the last few days."

"That's the point, Dr. Atkins," Grant interrupted. "That's why I'm sure you will want to help us. And once you see how important it is, you will be 100 percent behind us."

Allen and Phil saw the futility of resisting and were about to acquiesce, but Ann followed up on a previous comment of Grant.

"You mentioned earlier that you could make it worth our while to fully support you. What did you mean?"

Grant chuckled softly. "Your bio was accurate, Mrs. Atkins. You don't miss many financial opportunities. If you agree to help us, we will help you realize the full commercial aspect of this system, while maintaining certain aspects that could give us an advantage in the intelligence area, of course."

"That's all? We don't need your help to commercialize this system, Mr. Grant." Ann's voice was tinged with sarcasm.

Phil and Allen quickly picked up on Ann's interest in compensation for their efforts so far. Phil stood up and started pacing as he began a legal-like summary of their help so far.

"I should point out that with our help, Mr. Harrows uncovered information far more valuable than even he anticipated. The defector probably gave him the name of every North Korean agent in Southeast Asia. Even more importantly, the defector identified a double agent in one of our own intelligence organizations."

Allen sat back in his chair and folded his arms. "Yes, and Mr. Harrows admitted he had tried to duplicate the lie-detector system and couldn't identify the key component that enabled it to function. We had ours operational in a matter of minutes."

"We were also able to do this under extremely stressful conditions," Ann reminded them.

Phil continued his summary. "As civilians, we also were subjected to several attacks by mercenaries, in peacetime conditions. Surely, by now, we have demonstrated our ability to ferret out the truth even under difficult circumstances."

Ann finished the summary on a financial note. "It would seem reasonable, then, to assume THE top U.S. intelligence agency has already determined the value of the information we helped you obtain so far, and even the potential value of the information you are now asking us to help you acquire."

Grant had been watching them take turns justifying some level of compensation with a smile on his face. He suddenly laughed.

"John Reems' initial assessment of you was dead on. He referred to you in his report as the 'A Team' after the old TV series." He leaned over and conferred with an aide for a moment. "Okay. We are also prepared to offer you a considerable amount of money for your cooperation." Grant paused. He had their complete attention. "If you fully cooperate, we will authorize an amount that will be split equally among you and placed in your bank accounts in the amount of nine million dollars. Tax free of course."

Grant sat back amused as he watched their reactions. Each was dealing with that statement in their own mind, and it was reflected in their expressions. Phil was thinking of the impact of a third of nine million dollars, tax-free, could have on Joanne and the baby. He couldn't keep a smile off his face. Ann was thinking that a lot of money could help fund a whole new line of medical devices. Allen was thinking of all the money he would have to "tinker" with. Ann and Allen stared at each other for a moment before she finally answered Grant.

"That sounds reasonable. When can we start?"

Everyone laughed.

"In a few days. But first, I know you lost all your belongings while helping Mr. Harrows, so we will assign you an assistant to help you replace them... at our expense of course."

Ann's interest was suddenly piqued. "You mentioned this assignment would be in a luxury resort. Should we buy clothes here in Tokyo, or wait until we get there?"

"You can continue wearing your current clothes until you arrive at a temporary destination that will allow you to adjust to the time change and obtain your replacement clothes, or we can outfit you temporarily with a change of clothes here until you can acquire what you need."

"Where is the temporary destination? Uh... just to see what kind of shopping might be available there."

"The temporary city is Paris."

"Paris!" they all said at once.

The assignment looked more appealing all the time.

"We'll have a short briefing this evening, so we'll send a clothier to your hotel room to start replacing your things."

Allen needed very few suits. His dress at work was casual, and he only required a more formal outfit when an important government official visited. The clothier had done an excellent job. It was easily the finest suit he had ever owned. He looked at his reflection in a door-length mirror in the bedroom of their luxurious suite. Not bad. Allen turned sideways and noticed his stomach seemed to overhang the belt a little. He sucked in his stomach and let it out with a sigh.

The hell with it. Years of excesses had taken its toll.

He walked into the living room when he heard a key in the door. Ann walked in wearing an extremely expensive, custom-tailored suit. Ann laughed as Allen ogled her for a moment then rushed over to hug and kiss her.

"Do nice clothes turn you on, Allen?"

"In your case it certainly does, Mrs. Atkins."

"You're not so bad yourself." She wriggled out of Allen's grasp and walked to a mirror.

Allen tried to grab her again, but she ducked and he missed.

"Hey! Don't wrinkle my suit. We have to go to a briefing in a few minutes, remember?"

"Of course I do."

Allen reached into his pocket. "By the way, I have something for you."

Ann was curious. "What is it?"

He took her left hand and slipped a small diamond ring on her finger. "This is the best I could do here," he said apologetically.

Ann stared at the ring for a moment and choked back a few tears. "I wish you had done that fifteen years ago." She put her arms around his neck, and they kissed passionately. Ann felt his hands wandering and let go of him, stepping back.

"Remember the briefing...."

She assured him with a smile when she saw his disappointment. "I'll make it up to you tonight."

Allen was watching her pose at various angles in front of the mirror when there was a knock on the door. He opened it and Phil entered in his new expensive suit.

"Well? How do I look?"

"You should have been a corporate lawyer. It looks good on you."

There was a knock on the door. Allen answered it and Mark Dunlevy entered. He surveyed the three new well-dressed agents.

"Well, that certainly is an improvement over your old clothes. I think they are ready for us. Are you ready to go?"

"Eager to begin, section leader!" replied Allen.

Dunlevy laughed. "Okay, secret agents, let's go."

Phil, Ann, Allen, and Dunlevy were waiting in a plush conference room for the briefing to begin. The lie detector system had been assembled on a side table, and Allen had performed a quick ‹systems› check on it to verify everything was functioning correctly, just in case they needed to give a demonstration of its capabilities.

Grant entered with an assistant and two Army Generals he didn't identify. Dunlevy leaned over and whispered to Allen, Ann, and Phil that the Generals were there merely as 'observers.' The assistant set up a slide projector and Grant picked up a laser pointer and began the briefing. He pointedly did not introduce the Generals and they did not volunteer their names. Grant's presentation included several relevant slides.

"In a few days, sensitive negotiations will begin at a secure site just outside Paris between representatives of the U.S. along with several key allies, and several governments and organizations that have sponsored terrorist activities in the past. The purpose of the negotiations is to seek a more peaceful future by attempting to address, and find solutions for, the political and economic problems that are at the heart of terrorism in Europe and the Middle East."

"The 'allies' delegation will be led by several high ranking U.S. administration and NATO officials, but more than a dozen-allied countries have agreed to send representatives to the meeting as well. Press coverage will be kept to a bare minimum, and the talks are expected to last approximately two weeks."

"We, the intelligence community, are not optimistic that these talks will lead to anything substantial, but we do believe there is a unique opportunity to gather crucial data regarding these terrorist-sponsoring governments and organizations. While we will openly support the negotiations, we have also been examining several new potential methods of gathering information during the meeting, including your lie detector system. We think it could be invaluable if the truth can be separated from the lies our delegation will have to sift through."

Grant stopped when he saw Phil shaking his head. "Do you have a question, Mr. Conley?"

"Something doesn't seem quite right here. It sounds like there are two separate U.S. government functions going on, and the participants in each are not aware of the other one. Am I right, Mr. Grant?"

"The direct answer to your question is yes."

"I would also guess that at some level of our government, certain officials know about both of these programs."

"You are correct. But I'm not at liberty to say at what level."

Phil smiled at him. "I really don't want to know that."

Grant cleared his throat. "Yes...." Allen and Ann were avoiding his stare, as he continued.

"The biggest problem we have is trying to outsmart ourselves. Several of our own people are assisting with the security, and we must figure out how we can obtain the information we need without anyone, especially them, knowing what we are doing or uncovering our plans."

Ann glanced up from her notes. "What would happen if we WERE discovered?"

"That would depend on who found out. If it's the U.S. or its allies, we can work something out and keep the matter quiet. However, if it's the terrorist-sponsoring countries or organization's security forces, that's quite another story. We probably could not stop them from revealing the truth to the world's press and breaking off the talks at our expense. A rather unpleasant event for our allies and the intelligence community."

Phil seemed to sense there was something Grant was not volunteering. "Is that all? Just an embarrassment and the ending of talks that you don't think will lead anywhere. I think there's more."

Grant struggled to find an answer Phil would accept. "If they discover you, their security forces could kidnap you before we would know anything happened."

"And?"

"They could, in effect, treat you as prisoners of war." He paused, considering how to tell them the rest. "Or... they simply could refuse to negotiate or even acknowledge they have you."

Phil was absently drumming his fingers on the table. "Not a pretty prospect, Mr. Grant. How would you get us out under those conditions?"

Grant didn't answer immediately, and Ann, Allen and Phil exchanged concerned expressions. He avoided their stares as he admitted the truth. "We would have some difficulty doing that."

Phil sat back in his chair. "I hate to sound negative, Mr. Grant, but this whole undertaking smacks of some 'black ops' type scenario. The kind where no agency officially knows anything about it, and if asked by some congressional oversight committee, all would deny it even existed. Several recent examples come to mind."

Grant remained uncharacteristically silent. The Generals shifted uncomfortably in their chairs, and even Dunlevy avoided their stares.

Allen rose from his chair, stood behind Ann, and put his hands on her shoulders. "Well, for my part, no response is the same as agreement, and that's not an acceptable answer. We're not spies, or even in the intelligence business. You can't expect us to risk capture by an enemy known for killing innocent civilians without some guarantee of retrieval."

Grant was vague. "We would do everything in our power to get you back."

Allen was not impressed. "I seem to recall that several civilians have spent many harsh years at the hands of these people, and the intelligence agencies were not able to do anything about it. When these groups decided they had gotten all the publicity they could out of it, then they released the captives to a local government."

"There is some risk. I can't deny that. But it is much more likely that nothing like that will happen."

Sarcasm crept into Ann's response. "So, in effect, you are asking us to risk our lives for some money in a bank account?"

Grant was not used to defending his plans. "You seem to be overlooking the purpose of this undertaking! With the kind of knowledge we hope to gain from this, we could do some serious damage to these organizations... perhaps even put an end to some of them. Think of the future and the innocent civilians lives you would be saving."

"I think that's the military's and the intelligence agencies' jobs," replied Allen, unconvinced. "As a government employee, you swore an oath to that effect, if I'm not mistaken. Something about preserve, protect, and defend?"

Grant stood up, visibly frustrated. "I had hoped I could appeal to your sense of justice. There are some bigger issues here than money. However, I can't force you to agree to help us."

Dunlevy handed Grant a folder. He opened it and looked at the contents for a moment. "It looks like we will have to pay for what we think the information is worth."

He pulled several pieces of paper from the folder, closed it, and tossed it on the table. He walked around the conference table and handed one to each of them. An account had been opened in each of their names in a Swiss bank with an initial deposit in the amount of three million dollars.

"Maybe that's enough incentive to balance the risk. I can't offer you anymore without a government oversight committee getting involved, and I don't want to do that."

They all looked at the deposit slips without saying anything. Ann gazed at the slip with her name on it. She had seen checks and deposits for large amounts before and had even written a few large checks herself but had never seen her name on a deposit or a check with anything approaching that amount. Allen sat back down and stared off into space. That amount of after-tax money was almost as much as he could realistically expect to earn in his present job until he retired.

Phil wondered if the slips were real. Even if they cooperated and everything went as planned, would these accounts still exist? Or what if something went wrong and they were captured and held as prisoners for years. Would they be able to survive? Even if they survived, would they be in such terrible condition that they couldn't enjoy the rest of their lives once they were released, even if they had that much money?

George Grant seemed to know what Phil was thinking from the various expressions that crossed his face.

"Life's a gamble, Mr. Conley. Some people win a lottery, and their lives turn out worse than if they had never won."

Ann put her hand on Allen's. "So... it boils down to money. I've spent most of my life trying to achieve the kind of security that this amount of money can bring. But to be honest, I don't know if I could survive as a prisoner for any significant amount of time. I'll go with whatever you decide, Allen."

Phil sat back in his chair. "I've spent most of my career putting criminals behind bars. I agree with Ann... I don't know if I could survive as a prisoner either, but I'm willing to go along with whatever you decide. We obviously can't do this without you, Allen."

Allen was suddenly angry. "You can't dump this decision on me. You're both adults and responsible for your decisions as well."

"We're not asking you to decide for us," said Phil. "I think we just said we would only agree to this if you did."

"It's the same thing!"

Grant interrupted. "This IS a big decision. Why don't we leave and let you talk about it among yourselves. I wish I could give you several days to make up your minds, but in order to get everything ready, we need your decision before midnight tonight. Please inform Mark Dunlevy of your decision."

The two army Generals left quietly as Grant and his assistant gathered up sensitive intelligence data, preparing to leave them alone. Grant paused at the door and looked back. "One other thing. If you agree to help, I'll do my best to see you're allowed to pursue commercial ventures with your system... without any government interference. In any case, I'll understand if you decide not to help us."

When Grant's assistant closed the door, Dunlevy pulled a chair next to Allen.

"There is something else I need to tell the three of you. A day or so ago, Roger Douglas escaped while he was being transferred to a maximum-security prison. We believe he had some inside help, as well as outside help."

When they didn't respond to his indirect warning, Dunlevy continued.

"The reason I'm telling you this is that he ranted frequently about getting even with the three of you for uncovering his deception. One psychologist who examined him declared him mentally unstable, and a potential menace to society."

Ann felt a headache coming on and started rubbing her forehead. "More wonderful news."

Allen was a little more upbeat. "I would think the probability of running into him again is pretty low. Isn't he afraid of being recaptured?"

"Undoubtedly, but the North Koreans will also be at this conference, so a little while ago, we notified the security forces that will be guarding the conferees of his escape, just in case." Dunlevy stood up to leave. "I normally wouldn't have mentioned this, but I felt compelled to let you know. Please let me know when you do reach a decision," he closed the door behind him.

Phil retrieved a coffeepot from a side table and put it on the conference table in front of them. He looked at his watch. "Four hours."

Allen and Ann stood up and hugged each other as Phil picked up his deposit slip to count all the zeros again in the initial deposit.

After four tough hours of examining every possibility, greed won.

Allen picked up a phone to call Dunlevy. "Mark? This is Allen. We've agreed to help. What do you want us to do now?"

CHAPTER 37

Dunlevy informed them he would be accompanying them to Paris. He escorted them directly to a waiting Citation Ten. Somehow, they weren't surprised to find a small bag with their old clothes and the lie detector system suitcases waiting for them. Dunlevy obtained some unusually comfortable sleeping bags and they all spent most of the trip to Paris asleep on the floor of the jet.

Grant lived up to his promises. The sun was just setting as they checked into a five-star hotel in the middle of Paris to rest for a day in preparation of the main event to be held somewhere in the vicinity. Almost as soon as Phil checked into his room, there was a knock on the door. He opened it for Dunlevy.

"I have a surprise for you, compliments of Mr. Grant" He stepped aside as Joanne walked into view. Joanne was wearing a designer maternity dress and a big smile. Phil was so surprised he didn't react for a few seconds. He rushed forward and started hugging and kissing her so relentlessly, she struggled to break free.

"Phil! It's really me, and I'm not going anywhere!"

Phil let her go. "I'm sorry, honey. It's just at times I didn't think I would ever see you again."

Mark Dunlevy was smiling as he closed the door and walked off.

Early the next morning Allen and Ann joined Phil and Joanne for an all-day shopping spree to replace the clothing that had been destroyed along with the helicopter. Allen insisted he accompany Ann when she and Joanne mentioned the need to shop for new lingerie.

Ann confided to Phil that she had never seen Allen so amorous. "I'll wear an expensive suit all the time if he continues like this.

Later that evening, Ann and Allen stopped by the Bell Captain's desk to seek his advice on a nice restaurant for dinner.

Ann was holding a guidebook when he looked up at them. "What's the best restaurant in Paris?"

The Bell Captain sniffed at their obvious ignorance. "Le Clarence, but you can't get a reservation there, Monsieur. You need to make a reservation at least six months in advance."

Allen took out a cell phone Dunlevy had given them and spoke with his assistant. A few moments later, the Bell Captain's phone rang. The restaurant was sending a limousine over for them... right away!

Allen and Ann smiled at his amazed expression.

Ann and Allen finished what they agreed was the best meal they had ever eaten. Allen handed the waiter a titanium credit card Dunlevy had also given them. «It tastes even better when someone else pays.»

Ann put her hand on Allen's. "Where now?"

"I was thinking about a cruise down the Seine and then turning in a little early." He grinned at her, and she laughed.

"Haven't we caught up, yet?"

CHAPTER 38

At precisely 6:30AM the next day, plain-clothes intelligence officers knocked on Ann and Allen's door and Phil and Joanne's door. Phil was reluctant to leave Joanne, but she insisted she would be all right.

A limousine without any insignia or official markings was waiting for them. The trip was fairly short as they left the city and entered a huge country estate. For this initial meeting, Allen and Phil were wearing conservative, yet expensively tailored suits. Ann was wearing a similarly tasteful dress. Their dress and demeanor elicited few second looks or unusual scrutiny as they passed the initial security checkpoints. Dunlevy met and escorted them down a hall to a small conference room and began their briefing on the security arrangements for the conference.

"As the host country, France has the primary responsibility for security at the conference. We have to coordinate everything we do through them. In reality, it would be almost impossible to try and bypass them. They probably already planned to have every room at the conference site bugged even before we began our discussions with them. We explained the system's operation to them and the nature of the intelligence we hope to gather and, of course, how we will share the information with them."

"So far they have been very cooperative. In fact, several DST people have expressed an interest in talking to the three of you."

Phil frowned. "DST?"

"Directorate for the Surveillance of the Territory, or something like that... sort of like our FBI."

Ann glanced up from her notes. "What's next, Mark?"

"Several of our people have been trying to figure out a way to use your system without drawing the attention of the Sûreté... the French national police, or the security teams being provided by our allies, or the limited security personnel being provided by the other side."

Allen had been munching in earnest on a plate piled high with French pastries when Ann kicked him gently under the table.

"If you can take a break from that, I'd just as soon get started."

Allen blushed as he stuffed the remainder of a croissant into his mouth.

"Sorry," he mumbled.

Ann picked her purse up and started for the door. "When does the conference start?"

The others stood up as Dunlevy replied, "In four days."

"Four days!" exclaimed Allen. "That doesn't give us much time."

Dunlevy opened the door for them. "Then we had better get started." He led them to a modest yet comfortable room in the estate that had been temporarily converted to a combination conference and planning room. A half dozen technicians were hard at work assembling and checking out a maze of equipment that would be installed throughout the conference site. Communications Specialist Wayne Robbins motioned them to sit down, introduced himself, and began the meeting.

"Mr. Dunlevy and I have spent hours discussing the operation of your lie-detector system and what changes we will have to make to enable it to work and not be detected by one of the most sophisticated security systems in existence."

He turned on a slide projector and began a detailed explanation of the multi-faceted system that was already operational at the site. Quite a bit of it was beyond Phil and Ann's comprehension and even Allen asked Robbins numerous questions to help them understand the system.

"The biggest problem is that there cannot be any form or type of broadcast communications, as they are listening to everything from subsonic transmissions to satellite communications, and everything in between. We might have a chance if we could broadcast only for a few moments at random times of the day, but we hope to monitor the conferees continuously. This rules out most broadcast methods as they would surely be picked up."

Allen was absently rubbing his chin. "What about encryption?"

"That would work for a while, but they still would come and ask what the hell we are doing!"

"What about band skipping?"

Robbins and Dunlevy stared at him with blank expressions. "I'm not sure what you mean, Doctor," Robbins replied.

"Broadcasting in encrypted bursts. The data at the end of each burst defines the next frequency for the transmitter and receiver. A computer picks the frequencies for the bursts with a random number generator. The only requirement for the receiver is to be programmed with the seed number for the random number generator."

Robbins seemed to have been caught off guard. "Uh... we don't have anything like that. Is that a NASA development?"

"I had the impression it was a NSA or CIA project. I don't know which agency would have developed it."

Robbins clarified it for them. "The CIA only deals with human intelligence, the NSA is responsible only for intelligence gathered by the interception of signals." He leaned over and spoke softly to Dunlevy so only the people seated around the table could hear. "I think someone's been holding out on us."

"I think I'll have a little conversation with Mr. Grant. Excuse me." Dunlevy bounced out of his chair and almost ran to a secure phone.

Allen rubbed his forehead as he stared at several blueprints of the wired portion of the security system that had already been installed.

"The options that are left present a big problem in getting the information to the computer and then back to someone in the main conference room who could advise the negotiators of the results."

"Exactly, Dr. Atkins. We have been discussing this problem for days and can only think of one solution... infrared transmissions via optical fibers."

Ann and Phil were hopelessly lost, but Allen sat up in his chair, uncharacteristically excited. "Yes!! That would do it." He then began a dissertation on the potential problems and some possible solutions that even Robbins had difficulty following. Robbins finally had to interrupt Allen's discourse.

"Dr. Atkins! I think we need to get the rest of the team in here and go over some of your suggestions. I'm not sure I follow everything you've said."

Ann laughed. "I'm glad I'm not the only one."

Robbins excused himself as Allen began pacing the small conference room, musing to himself out loud. Phil whispered to Ann.

"I don't even know what they are talking about. Do you think they will even need us?"

Ann smiled and nodded. "I'm sure we wouldn't be here if they didn't need us. Even our government doesn't give out that much money for nothing."

Ann observed a wry smile on Dunlevy's face when he returned to the conference table. "Did you get in touch with Mr. Grant?"

Dunlevy leaned over and whispered. "He went ballistic. He's going to go ream someone out at the NSA."

Soon after Allen and the team finalized the hardware and techniques they would need to penetrate the security system, Ann was pressed into service helping the team design the hardware and interface to the system. She was amazed at the resources that had been made available to them. Almost every system component was available in a superior, yet miniaturized version. She wondered if some of the equipment they were using was developed for the space program. Most of it was surely classified.

Dunlevy reported that Grant had confirmed the existence of the band-skipping equipment at the NSA, but they felt it was too experimental in nature to be relied on. Even so, Grant felt compelled to pass his frustration at their lack of cooperation up the chain of command.

Several members of the French secret police became actively involved in the construction of the prototype. Marie Franchette, a DST Communications Specialist, in particular worked closely with Allen during the initial testing of system components. Even Ann couldn't help but notice Marie's resemblance to Janet Turner, and she kept a watchful eye on Allen to ensure he stayed focused on the new system. At least that's what she told herself.

In less than two days, the design team had completed a prototype system. It was quickly made ready for a trial run and demonstration to several members of the French DST, and several senior members of the intelligence agencies sponsoring their efforts, none of whom were introduced to Ann, Allen or Phil. Grant called the meeting to order.

Allen began the presentation by reviewing the basic system's operations and components and then reviewed the changes required to make the system invisible to the security system in place at the meeting site.

"The basic change in the system hardware is the method used to transmit the data gathered by the sensors to the laptop, and then providing a means of getting the results back to our conferees without being detected."

"As you know, all 'over the air' type wireless broadcasting methods could not be used as virtually all regular broadcast frequencies of the electromagnetic spectrum are being monitored. This necessitated converting to a totally optical method. Due to the availability of components, we chose the infrared region to carry the data."

"In cooperation with the French DST, members of our design team were able to place two infrared cameras in the main conference room and one in each of the several small conference rooms nearby, before the main security system was placed in operation. These cameras appear to be identical to regular surveillance cameras. And in fact, they do carry a regular TV signal back to the main security station."

"They also carry an infrared signal via an optical fiber to a small room in the basement, where our system will be installed. These infrared signals will carry the facial blood flow images that the system uses along with the voice stress analysis and breath composition information. A small, but powerful carbon dioxide laser is located in the base of the camera and can be targeted on an individual in the room, independently of the infrared camera. The system operator uses the infrared camera to align the laser."

There were no questions from DST representatives or the senior intelligence officers, so Allen continued. "There are several microphones in the conference room, and we have installed adapters on them so that they transmit voice as an infrared signal over an optical fiber to the lie detector system. All of this hardware just described will allow all the necessary information to be gathered by the system, but we still needed a mechanism to get the results back to our conferees."

"We have installed infrared transmitters on the surveillance cameras that will broadcast the return information into the room. Members of the design team will take turns acting as an aide to one of our conferees. He or she will have special glasses that pick up the transmissions and convert them to 'advisory' statements that can be passed to their conferee, virtually in real-time."

One of the senior officers got Allen's attention. "What is the nature of the advisory statement, Dr. Atkins."

"Basically a probability number from one to one hundred percent that the last statement or answer was true or not. We have prepared a small computer animation on the overall operation of the system."

The lights in the small conference room were dimmed and a computer-driven animation was projected onto a screen. The visitors watched as the animation revealed how the various signals were gathered from the targeted conferees in the large conference room, analyzed by the system and Sherlock and then retransmitted back to the conference room. The results were then picked up by a member of the design team who revealed the results to his contact on the conferee team. The visitors were very impressed with the concept.

Grant addressed the members in response to a question on how the system would be brought into the building past the security system and guards.

"Several members of the design team will act as news media, and in particular, a TV news crew representing several military organizations from Southeast Asia. They will carry in actual working equipment, including TV cameras and tape-editing equipment that will be inspected by conference security. The lie-detector components have been buried inside this equipment, thanks to one member of the design team who has extensive knowledge of this system and with building prototypes. She also happens to be Dr. Atkins' wife, Ann." Ann stood up and the visitors acknowledged her accomplishments with applause.

Grant also acknowledged Phil. "Phil Conley will be the aide to the conferees. He is a lawyer and also has extensive knowledge of the system." The visitors applauded politely for Phil.

When a member of the DST asked who would monitor the system in the basement, Grant informed him Allen and Ann would take turns doing that. A general, who had not been named, asked one final question.

"Has the overall system been tested?"

"Yes!" replied Allen. "For several hours. The results were even better than we hoped for. Would you like to see the system in operation?"

The members of the DST and intelligence agencies were eager participants in a final test of the system.

CHAPTER 39

A thick fog hid the sun's rise on the day before the negotiations were to begin. A «press» van glided to a stop at a side gate to what Ann, Allen and Phil at first thought was Versailles. They soon realized it was one of several summer homes of the former royal family of France. The gate proved only the first of several security stations the team would have to pass. Two technicians, Ann, and Allen exited the van as ordered while it was searched thoroughly for hidden weapons and bombs. Another security guard swiped their security passes through a scanner and verified their access permissions. They reboarded the van and pulled up to a trade entrance to unload their gear. Several guards watched them unload the camera cases and film-editing equipment without comment.

Guards entrusted with the castle's interior security followed them to a "pressroom" already set up with a bank of telephones, video equipment, and dozens of personal computers tied together with a local area network and connected to a high speed link to the Internet.

The guards gradually drifted away as the team pulled their TV gear out of the aluminum cases and began assembling various sub-sections. Marie Franchette and her two fellow DST agents were well known to the internal security forces and they were able to pass almost at will through the various security stations inside the estate. They soon struck up a casual conversation with Ann and Allen in the pressroom, and quietly slipped them French DST security passes.

The internal security forces were quickly coming under great stress as lower level delegates, aides to the conferees, and their security personnel arrived in large numbers to make last-minute preparations for their leaders. In the resulting confusion of orders and requests in multiple languages from the conferee's aides, it was relatively easy for Marie, Ann and Allen to leave the pressroom together. They were generally ignored as they carried several small aluminum suitcases through the trades area, down a stairwell crowded with service personnel, and through an underground maze of storage and equipment rooms. Marie, Ann and Allen passed numerous service and trade personnel too busy with their own tasks to even notice them.

Hak Suhendra was also on a mission, but peace wasn't one of its goals. He had barely managed to hang onto his position with the North Korean intelligence agency following the three disastrous attempts at recovering or eliminating Kim Sang. He even managed to convince his supervisor to allow him to accompany the delegation to the conference. He tried to describe the lie-detector system to them, and why he thought the CIA would probably try to use it or something like it at the conference. He finally gave up on that, but he did convince them he should lead the North Korean intelligence gathering team that would accompany the regular delegation.

He also arrived a day or so early with the other aides to search for the lie-detector system. His credentials were real, and he easily passed through the tight security system without drawing any attention to himself. The head of the North Korean delegation did not like Suhendra or trust him completely. He assigned one of his own aides to shadow Suhendra and report back to him on his activities. Suhendra quickly picked up on his 'tail' and confronted him in a small conference room. They argued in Korean in full view of the regular security personnel, confident no one would have a clue as to the nature of their argument.

Marie, Ann and Allen also passed several security personnel as they searched an enormous basement under the country estate and finally found a small room marked ‹Electrical Equipment Room - Authorized Personnel Only.› Allen's key opened the door and they quickly entered. They soon found their target, a specific electrical panel located among a long line of similar panels marked ‹Danger - Do Not Open Without Authorization.› Allen fished a smaller key from his shirt pocket and opened the panel. He was relieved to find 30 or so fiber optic cables neatly bundled together, labeled, and terminated with communication cable connectors.

Together, Marie and Allen dragged a table under the panel and Ann opened the suitcases they had accompanied all the way around the world. They quickly assembled the lie detector system and connected the fiber optic cables to a specially designed fiber optic interface box. Allen crossed his fingers as he turned the power on and waited for the system to boot up. Ann was confident the system would work and pulled a coffee thermos from the suitcase. She poured herself a cup of coffee as Allen initialized Sherlock and the Artificial Intelligence software. Marie assumed a guarding position near the door, ever alert for trouble.

Several windows opened on the flat panel TV monitor and a message appeared on the screen from Sherlock. Allen shook his head. He always wants to bullshit with me. Allen ignored Sherlock and checked out the video feeds from the various dual mode TV cameras that had been placed around the main conference room and several smaller conference rooms nearby. Everything seemed functional. Allen pulled a set of headphones on when he saw the voice stress analysis window open. Two minor aides from one of the terrorist sponsoring organizations were arguing in one of the small conference rooms loudly enough for the VSA to attempt to analyze their conversation. They were speaking in Korean but Allen could sense their own conference preparations were not going well. Allen decided to record the conversation for playback later to their translators.

On the day the conference opened, the sun broke through a persistent cloud cover. After weeks of dark and gloomy days, many conferees, and some of the press, took it to be a symbolic sign that the conference would succeed.

Phil accompanied one of the allies' negotiators, pretending to act as his aide. This enabled Phil to whisper the results of the lie-detector analyses in his ear. Phil's goal was to remain as inconspicuous as possible and to avoid the attention of the opposing team as well.

Phil watched expectantly, as news media from all over the globe took pictures and video of the opening remarks. His eyes finally found Allen and Ann pretending to supervise a videotaping of the proceedings along with scores of others. He pretended to ignore them while snatching an occasional glance in their direction.

When the opening ceremonies were concluded, the members of the press were quickly shuttled out and herded into the designated pressroom.

Reporters and camera crews could make periodic reports back to their base operations and were free to roam about the huge complex, but could not enter the negotiation area. News releases were also periodically distributed in the pressroom.

Marie quickly joined Allen and Ann as they passed numerous security personnel and slipped quickly and quietly into the small basement room to monitor the lie-detector equipment.

Roger Douglas had lifted pictures of Allen, Ann and Phil from a security folder and mailed copies to Hak Suhendra. Suhendra searched the complex thoroughly but could not find them, or the lie detector, or anything out of the ordinary. He couldn›t be as thorough as he would have liked, however, as he had to be careful to avoid suspicion. He casually surveyed the crowd of reporters and conferees and, for a second, thought he recognized the man that had almost killed him in Thailand, but he couldn›t be sure. His memory of the attack was still fuzzy, and he had not associated it with the picture of Allen.

His gaze was focused primarily on the civilians in the crowd, and he tended to ignore the press corps in the room. He completely missed Ann as he had never really paid much attention to her photo. Douglas hadn't even told him why she was in Thailand with Allen. He probably wouldn't have recognized her anyway.

When the press was ushered out of the room, Suhendra stayed long enough to study the faces of the conferees and their aides. He exchanged glances with Phil and even happened to be standing a few feet from him. He had only glanced in Phil's direction not expecting to find them in the middle of the conferees. When the conference started in earnest, he slowly left the room to continue his search, certain Allen and his lie detector were somewhere nearby.

Allen had earphones on and was listening intently to the conference when Dunlevy entered the power distribution room. Ann stood up from a chair where she had been reading one of her novels to greet him. She put a finger to her lips and they stepped out into the hall and she closed the door. She saw Dunlevy holding a folder.

"How's it going?"

"No big revelations so far."

"Do you remember that conversation Allen recorded yesterday?"

She nodded as he opened the folder and handed her a transcript of the conversation. "What was it about?"

"Apparently one of the aides to the North Korean delegation was keeping an eye on one of their security men for some reason. The person being followed was Hak Suhendra, and he didn't like it at all."

When Dunlevy saw her blank expression, he continued. "Suhendra was Roger Douglas' contact at the North Korean intelligence agency."

Her eyes grew wide. "The same Roger Douglas that swore revenge on us?"

He nodded.

"What are we going to do about him? Do you think Douglas is here as well?"

"Unfortunately, Suhendra is here as part of an official delegation, and it will be extremely difficult to arrest him unless he commits a crime. I doubt Douglas will show his face around here. He knows we're looking for him. But just in case someone starts poking around, we've stationed some extra guards near here."

Hak Suhendra had already poked around and was surprised when two French security guards stopped him as he tried to enter a storage area in the basement. He pretended to be lost and left quickly, cursing loudly at them in Korean. The guards stared after him for a moment and returned to their posts. He was determined to find out why they were guarding a storage area. A report was filed a few days later with the head of French security, describing the theft of a South Korean Colonel's uniform from a room in the compound's sleeping quarters. It didn't receive much attention given all the other security issues at the conference.

A week of intense negotiations passed without much progress. This was not unusual for two sides as far apart to simply put their position into perspective. The lie detector system had verified that, for the most part, the conferees for both sides were telling the truth, and the differences were mainly a different way of viewing the same set of facts.

Phil rarely had to advise his conferee of any lies or deceptions, and several senior intelligence officials were beginning to doubt the system was working. The team held a late-night huddle with Major Dunlevy to review the week's results.

"I'm not surprised they feel that way. I would have thought we would find more than we have."

"Don't take this as criticism, Dr. Atkins. They are simply worried the system may not be working and we may miss something when it does finally occur."

"I'm not an expert, but it seems to me the negotiations are starting to make progress. Perhaps when we get to the final negotiations, we'll find something worth all this effort."

"Perhaps you're right, Mr. Conley. I just wanted to pass along this along, in case you hear a slightly different story from one of Grant's assistants."

Ann appreciated his candidness. "We understand."

Allen couldn't forget Douglas's promise to extract revenge. "Any word on that guy Hak, or on Douglas?"

«Hak Suhendra is still under surveillance. He hasn›t tried anything unusual. Most of the time, he seems to wander aimlessly around for no purpose. We haven›t seen or heard anything about Douglas. He must be in hiding or waiting for the right moment to surprise us.»

Ann felt a sudden chill and shivered. "I hope not."

Allen saw her reaction and put his arms around her.

Hak Suhendra wasn›t accustomed to wearing a uniform, especially a South Korean Colonel's uniform. There was considerable risk in dressing as a military officer of another government, particularly South Korea, and Suhendra was determined to avoid detection. It had taken several days to have the uniform altered and even then, it just didn't seem to fit very well. At the end of the day's proceedings, he waited patiently as the conferees and their aides left. He had stashed the uniform in a utility closet in one of the bathrooms. He changed quickly, ever watchful for the security personnel just coming onto the evening shift. The building was quiet as he made his way down the stairs to the basement. He knew a direct tact was best and walked briskly toward an internal security guard holding an automatic rifle in a ready position. The guard saw his uniform and challenged him in French. "Excuse me sir. This is a restricted area. I'll have to see some identification and an authorization for you to enter."

Fortunately, Suhendra spoke French, and he flashed a newly minted identification badge and a forged security pass in front of the guard. The guard backed up for him to pass.

"Thank you, sir."

Suhendra nodded and walked quickly down the hall. He glanced back to be sure the guard wasn't watching him and began silently to test the doors. Some were locked, but the door to the tenth one he tried was open and the light in the room was on. He quickly ducked inside and found a janitor emptying several trashcans filled with computer printouts into a large waste bin on wheels. There was a 'Confidential - Burn Only' sign on the side of the bin. The janitor glanced at him and nodded. When he turned away, Hak picked a printout out of the bin and thumbed through it. A smile slowly appeared on his face as he tossed the printout back into the bin. He pulled out a secure digital cell phone and keyed in a telephone number in Paris.

He chuckled to himself. This should put him back in good standing with the agency... perhaps even get him a promotion. He also could use a vacation, and Paris would be nice. He turned on the computer as a familiar voice answered his call. "Hello, Roger? What do you want me to do?"

CHAPTER 40

Negotiations were suspended for the weekend to give the conferees a brief rest. Joanne had been planning the weekend for almost a week, and Phil was so happy to be away from the negotiations to want to change anything. They disappeared into the French countryside almost immediately.

Ann and Allen spent a restful, romantic weekend attending a play and dining in elegant restaurants. Somehow, these diversions made the whole time they were in Paris seem like a dream or a play they were acting out and helped block out the reality and danger of intelligence gathering.

Allen sat on the bed watching Ann open the balcony curtains and door and walk around their luxurious suite, turning off the lights. A partial moon peeked from behind some fast-moving clouds, furnishing a meager amount of light in the room. He grabbed her as she neared the bed and pulled her on top of him. A cool breeze blew in the open door and Ann pulled the sheet over them. A shadow passed over them drawing her attention away from Allen, but he was too busy to notice as the memory of the first time he had seen Ann in her nicely tailored suit still provided the fuel for the desire he felt for her.

"Allen, did you see something?" she whispered in his ear.

"Only you, dear."

"No, I mean... I thought I saw someone on the balcony."

Allen was out of the bed in a flash. He needed a weapon. He unscrewed the shade from the lamp and pulled the cord from the wall socket. He put his hand on Ann's mouth and dragged her off the bed to the floor. The only light in the room was filtered by the passing clouds.

Allen suddenly realized the suite entry door was open. Ann and Allen instinctively jumped when Roger Douglas turned on an overhead light. He was dressed in black and was aiming an automatic pistol at them. Behind him, were three North Koreans dressed in black outfits with hoods covering most of their heads. They were holding automatic weapons as well.

"Well, I'm sorry to disturb you both at this late hour, but I need to talk to you, Dr. Atkins." He motioned to them with the pistol to stand up and move to the living room.

Ann wrapped a sheet around her, and Allen quickly pulled on his underwear and pants. They walked past Douglas to the living room and sat down. The Koreans with Douglas searched the apartment quickly and took up guarding positions.

"What do you want?" demanded Allen.

Douglas motioned to the Koreans, and they lowered their weapons. "First, I'd like your wife to get dressed and accompany these men to a location where she will be well treated. Then we can talk."

Allen knew how frightened she was and hugged her. "Allen, I don't want to leave without you," she whispered.

"There's nothing I can do right now. There are too many guns, and I can't take the chance of you getting hurt. You'll have to go with them. I promise I'll do whatever it takes to get you back."

Ann was trying not to cry as Allen wiped her tears away. He nodded to her, and she gathered her clothes and went into the bathroom to dress. Douglas never took his gun off Allen as he finished dressing and sat down on a sofa.

"What do you want?" Allen repeated, fighting an impulse to try and get the jump on Douglas and strangle him.

"I need your help."

Allen's puzzlement showed, and Douglas smiled.

Ann walked into the room and Allen stood up and hugged and kissed her. "I'll be all right, Allen," she whispered in his ear. "Just be careful. Remember what Dunlevy said about him... he's insane."

Allen let go of her, and she walked to the door. One of the Koreans opened it, looked up and down the hallway and walked quickly out, leading the way. The last one out closed the door and Allen heard them walk off.

Douglas stood up and started pacing the floor. "I wasn't really a double agent. Only one other person within the agency knew I was acting as one, to give the Koreans false information."

A triple agent? "Why didn't you tell Dunlevy or Grant?"

"I tried to, but I don't have any proof. The only person in the agency that knew what I was doing, died of a heart attack two months ago. My problem now is that I had to give the North Koreans some factual intelligence to convince them I was for real. It wasn't anything that important, and nothing that endangered lives or anything, but it was real. Dunlevy is a real tough guy. Once he makes up his mind, you can't change it. Grant is even worse. They are both convinced I'm a traitor."

"What does all of this have to do with me or my wife?"

"I figured out the CIA would never pass up on an opportunity to use an invention like yours to gather some intelligence at a meeting like this, so I knew you'd be here. When you find out the North Koreans have no intention of making a deal, I need you to give me the proof and I'll give it to Dunlevy instead of the North Koreans to show him I'm no traitor."

"What about my wife?"

"I'm sorry about that, but I had to make it look to the North Koreans like I'm holding her to make you cooperate and give me intelligence they can use. I promise you, she will not be harmed in any way, and will be treated well. I have no interest in seeing her harmed."

Allen sat thinking for a while, considering his options. What if this were all a big lie? What if Douglas was just making this up to throw him off? What if he gave Douglas the information he wanted and he passed it to the North Koreans? How would he ever get Ann back?

"What if you're lying now?"

"What can I say? How can I prove that what I'm saying is real?"

Douglas was still pointing his gun at him.

"If you are for real, why don't you lower that gun?"

Douglas chuckled. "I'm in the intelligence area, remember. I know you are a martial arts expert and what you did to Harrows. If I lowered this gun, you would probably kill me in some weird and painful way."

257

Allen was still sorry he had pounded Harrows so thoroughly. "I really didn't mean to hit him that hard."

"I heard he still has headaches."

Allen suddenly felt very tired. A memory of a martial arts tournament flashed though his consciousness and the sight of an unconscious competitor being carried out of the sports arena. Allen sat back on the sofa and closed his eyes.

Douglas studied him and slowly lowered his gun.

"Dr. Atkins?"

Allen opened his eyes and stared at Douglas. "What?"

"Will you help me?"

"I'm sorry, I just don't believe you."

"Why don't you use your lie detector on me and see if I'm telling the truth or not?"

Allen hadn't thought of that. He wondered if Douglas could fool Sherlock. He had plenty of time to figure out a way to do it, and he knew a lot about the system. But there didn't seem to be any other way. "We could never get in there at night. Security is too tight."

"Surely you know someone who could help get you in, or maybe even bring the system here."

Marie! Maybe she would help if he explained the situation. "I may know someone."

Douglas handed him a phone.

Marie Franchette was more than a little surprised and somewhat pleased when Allen called her. Her emotions were soon replaced by concern as Allen described Ann's situation. «Of course, I can help," she replied. "Meet me in the lobby in fifteen minutes."

As they stood up to leave, Douglas seemed uncertain about something. He then flipped the gun over and held it out handle first, to Allen. Allen was so surprised at first he didn't react.

"I'm taking a pretty big chance on this, but I need you on my side, so... here's your chance to kill me."

"I don't want to kill you, I just want my wife back."

"I'm afraid I can't help you there. They won't give her back to you, or me, or anyone without something in exchange."

"But you know where they are taking her!"

"It doesn't matter, they would kill her if we tried."

Douglas was still holding the gun out to him. "Keep it. I don't like guns."

"After you, Doctor." Douglas put the gun in a shoulder holster as he opened the door for Allen.

Allen and Douglas were alone in the huge lobby as an antique mantle clock struck 1:00AM. They stood up immediately when Marie joined them. Marie was immediately suspicious of Douglas, and she told him so. «I don›t believe you, Mr. Douglas.»

Douglas merely shrugged it off. "Believe what you will. But give me the benefit of the doubt if I'm able to pass the lie detector."

Marie mulled that over until Allen walked to the lobby door. "Shall we go?"

A private, black sedan slowly pulled up to the main gate of the estate and a window rolled down. The night guard immediately recognized Marie as the communications team leader for the DST.

"Good evening, Miss Franchette."

"Good evening, Henri. How are you doing?"

Henri began a little small talk until he swept the beam of his flashlight in the car and illuminated Allen and Douglas. Both were wearing the same security badges as Marie.

"They're with me."

Henri nodded, backed up, and pressed a button to raise the gate. He saw the window of the sedan roll up as they drove off and a small white paper flutter to the ground. He picked it up and shined his flashlight on it. It simply had a number on it. He knew the police sometimes communicated with the internal security forces via codes and picked up a phone to call his supervisor.

Henri read the number to his supervisor several times who concluded Marie must have transposed two numbers. Just in case, his supervisor called Mark Dunlevy's hotel room.

CHAPTER 41

Allen, Marie, and Douglas quickly entered the basement area. The guard on duty recognized Marie but pointed his weapon at Allen and Douglas.

"They're with me." Marie stood guard by the door as Allen and Douglas hurried into the equipment room. Allen had the system operational in a few minutes. Sherlock recognized Allen and Douglas from their voiceprints as they were discussing the test.

'Allen?' flashed on the screen and Allen turned on the laptop speakers.

"Good Morning, Allen, Mr. Douglas. What are you doing here at 1:30AM in the morning? Has the conference restarted?"

Douglas shook his head. "I wish I had one of these a few years ago. I would probably be giving Grant some orders by now." He turned to the laptop. "No, but we need you to open a new file for us."

"What is the nature of the file?"

"Interrogation and fact finding."

"File opened."

Allen motioned to Douglas. "Sit down on that chair." He pulled a miniature version of the infrared camera, laser, and spectrophotometer out of a briefcase and set it up, aiming the components at Douglas. These components were used to help troubleshoot the system. If a problem developed, these components could isolate the problem to the equipment in the room or the monitoring equipment in the main conference room.

Several windows opened on the computer screen as the Artificial Intelligence system became active. Sherlock needed just one more fact to begin. "Who is seated in the interrogation chair?"

Allen finished typing in the last command. "Roger Douglas"

Douglas had carelessly laid his gun on the table next to the laptop and Allen quickly slipped it into his pants just in case Douglas turned out to be a traitor. He decided to skip the usual basic questions used to establish a reference point and then hit Douglas with the key question.

"Are you a spy for the North Koreans?"

Douglas stared at him and replied calmly, "No."

"The last response of Roger Douglas had a probability of 90%" intoned Sherlock.

"See, I told you so." He chided Allen.

"Does your loyalty lay with the United States?"

"Yes, of course."

"The last response of Roger Douglas had a probability of 95%."

"If I gave you some intelligence data that has been gathered here, would you give it to the United States or some other country?"

"To the U.S." he replied without hesitation. Allen was staring at the monitor and didn't see Douglas turn the laptop speakers off. "You can ask me all the questions you like, Doctor, but the time is slipping away and I need to get my life back, along with your wife."

Allen hesitated, then gave the gun back to Douglas who put it in a shoulder holster under his jacket.

"What do you need?"

"Proof that the North Koreans and their allies have no intention of reaching an agreement here. It's just an opportunity for them to gather intelligence, just as it was for Grant."

Something about Douglas's offhand comment rang true to Allen. He picked up a thick folder of computer paper that had been punched and bound into a folder. He thumbed through it and found what he was looking for. He handed it to Douglas who sat down reading it as quickly as he could. He thumbed through the rest of the folder and suddenly stopped, looking at Allen.

"This is excellent, just what I need."

"Is it enough to get Ann back?"

Douglas nodded and stood up. "Where are Grant and Dunlevy staying?"

"At a hotel near here," replied Allen. He wrote the address down and handed it to Douglas.

"As soon as I clear my name, I'll go and get your wife." He glanced at the printout. "I'll have to edit this some, of course, or Grant will think I'm a traitor for sure."

He smiled at Allen and shook his hand. "You've been a great help, Dr. Atkins. Well, I'm off then."

He turned and stopped at the door to look back at Allen. "You can't begin to know how important this is." He waved and walked out.

Allen sat down wondering what he meant. Maybe he should replay the interrogation again.

"Sherlock?"

Sherlock didn't reply and Allen glanced quickly over the system. He found the speakers off and turned them on. *I don't remember doing that.*

"Sherlock?"

"Yes, Allen?"

"Did Roger Douglas tell a lie?"

"Yes. The last three responses were lies."

"What!" Allen began to scan the captured dialogue. How could those responses thanking him for his help be lies? Allen stared at the laptop wondering what was going on. He quickly closed all the applications and frantically searched everything for a possible problem. He finally found it. Someone had reversed the arithmetic sign of all the constants used by the Artificial Intelligence system to determine truth from lies. Every positive constant was now negative, and every negative constant was now positive. What would that mean?

"That would mean every truth would appear to be a lie and every lie would appear to be the truth." He muttered to himself.

A sinking feeling began to set in as Allen recalled how sensitive the data was that he had just given Douglas. Maybe he could catch him! Allen jumped up and ran down the hallway to the basement door. He saw the guard lying face down by the door and Douglas's gun next to him. Allen rolled the guard over. He had been shot but was still alive. Why would Douglas shoot the guard? He suddenly remembered that he had held that gun and Douglas had been wearing gloves. As he stared at the gun, he realized his fingerprints were still on it! If the guard died, all the evidence would point to him. He thought about wiping his fingerprints off the gun, but he couldn't afford to stop, or Douglas might get away. He could only hope the guard would live long enough to identify Douglas.

He ran up the stairs and out the trades door just in time to see Dunlevy and a contingent of the French Sûreté blocking the main gate. Douglas was speeding toward them in his car. Douglas bent over low and tried to blast through them, but the Sûreté had rolled a concrete barricade in front of the gate that Douglas couldn't see in the dark. The car smashed into it, and the gas tank exploded. Allen could see the interior of the car burst into flames. He ran to help, but the heat was so intense, he could only watch along with the other onlookers, as the flames consumed the car and driver.

Dunlevy was surprised to see Allen. "What are you doing here, Dr. Atkins?"

With the only apparent link to his wife gone, Allen slumped against the wall next to the gate.

"Are you all right?"

"No, I'm afraid not."

It only took a few minutes to explain everything to Dunlevy who listened at first in amazement, then with an obvious concern for Ann and Allen.

"What can we do?" Allen asked him desperately.

"I don't know. Maybe Mr. Grant has some ideas."

CHAPTER 42

«How can anything be accomplished when these civilians run amuck?» Grant thundered to Dunlevy, Allen, and his aides. His aides tried to become invisible as Grant stood up from his desk and began to pace back and forth.

He glared at Allen. "I told you to come to me if anything out of the ordinary happened, didn't I?"

Allen nodded wearily. "Why can't we just ask the North Koreans to give her back? She can't be worth anything to them."

"Give her back! Why should they? They probably don't even know what Douglas was up to."

"Someone at some level in the North Korean intelligence must know."

Grant sat down at his desk and rubbed his eyes. "Maybe some sort of a discreet inquiry though political channels...." he mused aloud.

"Is there anyone here who could start asking them?"

Grant shook his head. "We're talking Assistant Secretary of State. This will take some time."

"We don't have a lot of time!" cried Allen. "She must be somewhere in Paris."

"We don't know that."

Allen felt a sudden surge of energy. He walked over and stood in front of Grant who looked at him questioningly.

"We must have someone they would be willing to trade for her."

Grant frowned. "We don't have that kind of pull. Remember this is still an unofficial activity."

Allen folded his arms, his anger rising. "You said earlier that someone at some level knew about the negotiations AND your efforts here."

Grant sat back in his chair and waved Allen off. "I also said, 'Don't ask who that is.'"

"I am asking."

"Forget it, Doctor. It's out of the question."

"I don't think so." There was an undertone to Allen's comment that caused Grant to exchange glances with Dunlevy.

"I hoped I would never have to do this, but I have a 'Get Out of Jail Free' card up my sleeve as well."

Grant's concern was starting to show. "What do you mean?"

"Do you remember that first briefing you gave us?"

Grant sat back thinking for a moment. "Yes?"

"I recorded it and sent a copy in an e-mail to a friend of mine via the Internet. I said in the note to send it to the media if I were to die unexpectedly or from unknown causes."

Grant jumped to his feet, towering over Allen. "YOU BETTER BE LYING, MISTER!"

Allen calmly reached into his pocket and pulled out a USB drive. "See for yourself. Here's a copy." He tossed it onto Grant's desk and sat down in a chair in front of the desk, drained of energy.

Grant's attention turned to Dunlevy. "Could he be telling the truth?"

Dunlevy had his head down hoping Grant would not confront him. He knew he couldn't get out of this one. "Dr. Atkins is a very resourceful man, sir. I have no doubt he could have done that. Do you want me to check the disk?"

Grant sat down hard in his chair staring off into space. "Do you know what you are asking?"

"I think I do."

Allen had indeed stirred up a hornet's nest. He was threatened more than once and offered all kinds of inducements to hand over the original disk and all the copies he had made. He refused to even talk about anything else but a trade for Ann. After 24 hours of almost continuous negotiations, Allen returned to his hotel room to rest, utterly exhausted. At 4:00AM, an assistant of Grant knocked on his door, waking him.

"Mr. Grant would like to see you now. Please take the elevator to the third floor." The assistant left abruptly. Allen was surprised that Grant had come to the hotel instead of just calling. *What's going on?*

When Allen got off the elevator, he was frisked by two agents and escorted to a small conference room, where Grant, Dunlevy, and another man were waiting. The third man was seated with his back to the door, but Allen saw the unhappy expressions on Grant and Dunlevy's faces and entered the room gingerly. Instead of his usual bellowing and fits of rage, Grant seemed remarkably calm as he waved Allen in. He wondered what had happened. The third man slowly stood up and turned to meet Allen. He held out his hand.

"Guy Dunbar, Dr. Atkins. Sorry we couldn't meet under happier circumstances. I've heard a lot about you and your lie-detector system."

Allen stared at Dunbar as he shook his hand. I know him. It suddenly came to him... CIA Assistant Director Guy Dunbar! Allen had seen an evening news clip of him at a press conference discussing the ongoing peace conference and its prospects for progress.

Allen nodded in reply and swallowed hard as he sat down near Dunbar.

"Let me begin by saying the North Koreans at first insisted they knew nothing about your wife. It took quite some effort to finally find someone in their intelligence agency willing to listen to our proposal for an exchange."

Allen was too weary to wait for the bottom line.

"Mr. Dunbar... have they agreed to a swap or not?"

"Yes. I can't tell you how difficult it was to find someone they wanted in exchange for your wife. It seems they finally realized the value of your system and hoped she knew enough to help them construct one."

"I don't want to know any more," replied Allen wearily.

CHAPTER 43

The sun was trying to break through a thick morning fog as Allen waited with Grant and Dunlevy on a rural road outside Paris. At 7:00AM sharp, a black sedan rounded a curve in the road ahead of them and slowly pulled to a stop 50 yards away. Allen felt his heart pounding as several North Koreans got out. He recognized one of them as one of the guards that had taken Ann away. He felt a deep sense of relief when Ann stepped out of the car. She looked extremely tired but in good physical condition.

Allen heard a car door behind him open and he glanced back as a wizened old man with white hair slowly shuffled past him. Ann started walking toward them and they passed without even glancing at each other. Ann ran the last few yards into Allen's waiting arms. He closed his eyes as he hugged her, thankful she was all right. The North Koreans held the sedan's door open for the old man who slowly bent over and climbed inside.

Allen gave a quizzical look at Dunlevy.

"Don't ask."

Allen didn't. He helped Ann back to their car where Grant was waiting. Allen helped her inside and sat close to her.

"You don't even have an idea what you've done...." Grant began.

"Yes, I do!" interrupted Allen. "And I don't give a damn about that old man and whatever you thought he was worth."

Grant started to reply but shut up when Allen pulled Ann against himself and started caressing her hair.

The conference ended without resolving any key issues, but both sides issued statements they thought an important dialogue had been started, and were hopeful that future negotiations could lead to some real solutions.

Grant and Guy Dunbar asked to see Phil, Ann and Allen one last time before they headed back to Paris for some well-deserved R&R before finally heading home.

Grant was truly appreciative of their contribution. "Even though this didn't turn out like we planned, I wanted to let you know that we did gain some valuable intelligence and wanted to thank you all."

Dunbar was somewhat apologetic. "Unfortunately, in light of what has happened here, it will be necessary for us to reclaim the payment you were promised. I hope there's no hard feelings."

Ann laughed. "Somehow, I didn't think I would ever see any of that money, anyway."

Phil reminded them of one of Grant's earlier comments. "Life's a gamble, as you said, Mr. Grant."

They shook hands and Grant and Dunlevy escorted them to a waiting limousine for a short ride to Paris.

Dunlevy wished them well and closed the door. He waved at them as the limousine pulled away.

Phil stared out the window at the passing countryside. "There's still a chance they won't try and stop the joint ventures with Adam."

"I think the chances of that are slim and none." Ann took a deep breath, exhaling slowly. "Oh, well. Easy come, easy go." She leaned against Allen, and he put his arm around her as the limousine exited the main gate of the estate and sped down a rural road toward Paris.

CHAPTER 44

Allen had been away from his job for almost two months. When he returned to his office, the mail was piled up in a U.S. Postal bin on a chair. He sat down heavily in his own comfortable chair and stared out the window. Maybe it all hadn›t even happened.

Despite everything else, he was extremely happy to be married to Ann. He stared at the small golden ring on his finger and made a mental note to look for a larger diamond for Ann's ring. He almost laughed out loud. *If I can afford it.*

A financial friend told Allen the purchase and funding of Sleuth Software by investors could take some time, especially when challenges to patents were involved. So, there was no timetable yet on when he could see some cash from the shares of his partner's sale to Adam Daniels and a group of investors.

He suddenly thought about the lie detector system and wondered if the first prototype was still operational. Friends and colleagues stopped him to welcome him back on the way to his lab. Nothing had really changed since he left. He pushed open the door to the small lab in the back and found a small cardboard box on the table in the middle. He opened it and found it to be full of literature for a Christian summer camp for kids. He picked up a flyer and glanced through it, wondering how they got his name and why they sent so many copies. He glanced down again at the stack of literature and saw a tiny bulge in the middle of the stack of brochures. He pulled some flyers out of the way and found a key. He held it up and examined it but it didn›t have a name or number on it.

He glanced at the brochure again and the name of the organization sponsoring the camp. "Christians In Action," he said aloud. He repeated the name a few times until something dawned on him. He ran to the nearest phone and dial Ann's number at work.

"Hi, honey!" she answered. It still seemed strange to hear Ann say that. "What's going on?"

"Ann, did you find a small package on your desk this morning."

"Yes, but I threw it away. It was full of some literature for a summer camp or something. Why?"

"Do you still have the box?"

"Yes, it's right here in my trashcan. What's going on?"

"Look under the literature."

There was a pause and then she replied. "There's an address on an index card."

"Read it to me!"

"It's kind of strange, but I think it's the International Airport here in town. Are you going to tell me what's going on or not?"

"I will soon. Good-bye, dear." Allen was punching in Phil's phone number almost as soon as he hung up with Ann.

"Hi, Allen. It seems strange being back at work, doesn't it."

"It sure does. Phil, did you get a small package with literature for a summer camp?"

"Yes, why?"

"Would you mind looking through the brochures and see if you find anything there in addition to the literature."

"Sure, just a moment." After a short pause he replied. "The only thing I see is the number 505 written on an index card. What's it all about, Allen?"

"I'm not sure, yet. But I'll call you later. Thanks, Phil."

Allen drove to the international airport as quickly as he could without getting a ticket. He ran through the terminals until he found locker 505. He was breathing hard from the exertion but held his breath as he inserted the key.

A large cardboard box completely filled the locker and Allen struggled to get it out. He carried it to a nearby bench and opened it. It was full of Styrofoam peanuts, but he fished out a miniature infrared camera, a miniature spectrophotometer, and the remaining pieces needed to make another state-of-the-art lie detector system. He marveled at the latest versions. They were even smaller than the versions they had used in Paris. He also found a note from George Grant.

Dear Dr. Atkins:

It took a whole lot longer to get permission for you to use this stuff than I thought, but it's yours. I'm sure you, Mr. Conley, and Mrs. Atkins will find a way to exploit the concept commercially.

Oh, by the way, there is only one string attached to this stuff (there's always a string). Through your use of it, you agree to help us one more time if we should ever need it.

Thanks again, and good luck,

George Grant
Vice President,
Carborundum Inspection and Analysis

Allen carefully placed everything back in the box and almost ran with it to the nearest phone to call Ann and Phil. He called Phil to volunteer use of the new lie detector equipment for his hung jury trial. Ann immediately declared a holiday at Stevens Equipment Company and catered a huge party so all the employees could help them share in their good fortune.

Phil and Joanne's daughter was born on the same day three new corporations became official.

THE END

www.ingramcontent.com/pod-product-compliance
Lightning Source LLC
Chambersburg PA
CBHW062121020426
42335CB00013B/1053